Florian Muhle, Indra Bock (eds.)
Communicative AI in (Inter-)Action

BiUP General

Florian Muhle works as a professor for communication studies with a focus on digital communication at Zeppelin Universität Friedrichshafen, Germany. He is also affiliated at the Virtual Observatory for the Study of Online Networks (VOSON) Lab at the Australian National University in Canberra. One of the focal points of his research lies in the theoretical and empirical examination of various forms of automated communication.

Indra Bock is a Ph.D. student at Bielefeld Graduate School in history and sociology and a research fellow at Zeppelin Universität Friedrichshafen in the 3B Bots Building Bridges project. Her background lies in qualitative social research and media sociology. Her work surrounds human-robot interaction as well as automated communication in Online Social Networks and its influence on political opinion formation.

Florian Muhle, Indra Bock (eds.)

Communicative AI in (Inter-)Action

Investigating Human-Machine Encounter
outside the Laboratory

[transcript]

Funded by the Deutsche Forschungsgemeinschaft (DFG, German Research Foundation) and by Bielefeld University's Publication Fund

Bibliographic information published by the Deutsche Nationalbibliothek
The Deutsche Nationalbibliothek lists this publication in the Deutsche Nationalbibliografie; detailed bibliographic data are available in the Internet at https://dnb.dnb.de/

First published in 2024 by Bielefeld University Press, Bielefeld
© **Florian Muhle, Indra Bock (eds.)**
An Imprint of transcript Verlag https://www.transcript-verlag.de/bielefeld-up

Cover layout: Maria Arndt, Bielefeld
Proofread: Wiley Editing Services
Printed by: Majuskel Medienproduktion GmbH, Wetzlar
https://doi.org/10.14361/9783839475010
Print-ISBN: 978-3-8376-7501-6
PDF-ISBN: 978-3-8394-7501-0

Contents

Part IV: Methodological Issues

Conclusion

Communicative AI in (Inter-)Action: An Introduction

Florian Muhle, Indra Bock

Abstract *In recent years, communicative artificial intelligence (AI) technologies such as social robots, embodied agents, and smart speakers have begun to leave universities' and tech companies' laboratories to enter different domains of everyday life, including private households, museums, care facilities, and other institutional settings. This new situation is not only a challenge for the technical artifacts themselves, which need to perform "in the wild," but also for research, as this anthology will show. This introduction aims at providing background information to the development of communicative AI, which will be outlined, as well as the associated methodological research challenges. Based on this, the anthology is presented in four sections: (1) social robots in (inter-)action, (2) embodied agents in (inter-)action, (3) smart speakers in (inter-)action, and (4) methodological issues. Within this structure, the individual contributions are briefly outlined as follows.*

1. From Information Processing to Communicative AI

In recent years, new kinds of interactive technologies such as social robots, embodied agents, and smart speakers have started to leave universities' and tech companies' research laboratories. These new types of communication technologies are intended to interact with humans directly in different domains of the social world. Social robots, for instance, can be found in care facilities and museums, while embodied agents inhabit (commercial) websites or virtual worlds, and smart speakers have already entered millions of private households. These technologies, which can be characterized as "communicative AI (artificial intelligence)"[1] are the most recent actualizations of an old dream of mankind's: the idea of creating autonomous artificial persons that are able to interact with and like humans. The roots of this idea go back to antiquity at least, as evidenced by the fact that the development of artificial

---

1   Andrea L. Guzman/ Seth C. Lewis, Artificial Intelligence and Communication: A Human–Machine Communication Research Agenda, in: New Media & Society 22 (1/2020), 70–86; Hendrik Kempt, Chatbots and the Domestication of AI. A Relational Approach, Cham 2020.

persons played an important role in ancient Greek mythology. Indeed, Pamela Mc-Corduck,[2] in her "Inquiry into the History and Prospects of Artificial Intelligence", writes that "perhaps the earliest examples of the urge to make artificial persons are the Greek gods." However, it was not just the Greek gods but also Greek scholars who attempted to build self-acting automata about 2,000 years ago. For example, Heron of Alexandria developed an automated theater that has become famous for its stage that opens and closes independently and its figures that move automatically.

Early automata and their successors in the following centuries were *mechanical* devices, but today's situation looks very different. Communicative AI technologies are the "children" of modern computer technology. However, in contrast to the traditional AI systems developed since the 1950s, as well as to contemporary machine learning technologies that operate as *information processing* systems, communicative AI is not intended to solve problems or conduct complex computing operations in place of human beings. What distinguishes communicative AI from conventional computer technologies and other forms of AI is the fact that communicative AI systems are used for communicative purposes.[3] They are developed to allow people to interact with machines in a "natural" and intuitive manner.[4] In this sense, the development of communicative AI reflects a paradigm shift in the development of technical systems. In contrast to traditional computers and AI systems, communicative AI technologies are not primarily considered to be tools or a medium for communication but are rather humanlike interaction partners who engage in communication and potentially also develop social relationships with their human counterparts.[5]

---

2   Pamela McCorduck, Machines Who Think. A Personal Inquiry into the History and Prospects of Artificial Intelligence, Natick 2004.

3   Kempt, Chatbots and the Domestication of AI. A Relational Approach, 3.

4   Florian Muhle/Indra Bock, Intuitive Interfaces? Interface Design and its Impact on Human-Robot Interaction, in: Mensch und Computer 2019 – Workshopband, Bonn 2019, 346–347; Ipke Wachsmuth, Embodied Cooperative Systems: From Tool to Partnership, in: Catrin Misselhorn (ed.), Collective Agency and Cooperation in Natural and Artificial Systems. Explanation, Implementation and Simulation, Cham et al. 2015, 63–79; Guzman/Lewis, Artificial Intelligence and Communication, 70–86.

5   Nuno Afonso/Rui Prada, Agents That Relate: Improving the Social Believability of Non-Player Characters in Role-Playing Games, in: Scott M. Stevens/Shirley J. Saldamarco (eds.), Entertainment Computing—ICEC 2008, vol. 5309, Berlin, Heidelberg 2009, 34–45; Cynthia L. Breazeal, Designing sociable robots, Cambridge 2002; Kerstin Dautenhahn, Socially intelligent robots: dimensions of human–robot interaction, in: Philosophical Transactions of the Royal Society B. Biological Sciences 362 (1480/2007b), 679–704; Rui Prada, Ana Paiva, Human-Agent Interaction: Challenges for Bringing Humans and Agents Together, in: HAIDM—3rd International Workshop on Human-Agent Interaction Design and Models held at AAMAS'2014—13th International Conference on Autonomous Agents and Multi-Agent Systems, Paris 2014; Shanyang Zhao, Humanoid social robots as a medium of communication, in: New Media & Society 8 (2006), 401–419.

This is exactly the reason why it makes sense to consider embodied agents, social robots, smart speakers, and the like as forms of *communicative* AI.

2. The Historical Development of Communicative AI

As communication scientists Guzman and Lewis state, "for more than 70 years, the study of artificial intelligence (AI) and the study of communication have proceeded along separate trajectories"[6]. Nevertheless, communication has played an important role in both the *theory* and *practice* of AI research since the beginning of the 'Artificial Intelligence' research program.[7] From the very beginning, there has been a controversial theoretical debate as to whether machines can be intelligent. Opinions differed on these questions—both among AI researchers and philosophers, who quickly entered into corresponding debates.

A central contribution to this debate, which still shapes it today, comes from the British mathematician Alan Turing, who also laid the foundations for the development of the modern computer.[8] In his famous essay 'Computing Machinery and Intelligence', he rejected the ontological question "Can machines think?", which was at the center of the discussion at the time, as unhelpful.[9] Instead, he suggested asking whether machines are capable of giving reliable answers in a question-and-answer game that are indistinguishable from human answers.

In order to answer this question, Turing proposed a test setting which he himself called 'imitation game'[10] and which is known today as Turing test. The test setup involves a computer (A) and a person (B), each connected to another person, the 'interrogator', who "stays in a room apart from the other two"[11]. The task of the interrogator, who knows that there is a human and a machine on the other side, is to find out who is who by asking clever questions on any topic. Person B has the task of answering as authentically as possible, while the computer is supposed to simulate being a human being. The decisive question in this game is whether the machine succeeds in 'passing' as a human being. As David Gunkel puts it,

> "it is this question, according to Turing, that *replaces* the initial and unfortunately ambiguous inquiry "Can machines think? Consequently, if a computer does in fact

---

6   Guzman/Lewis, Artificial Intelligence and Communication, 71.

7   David J. Gunkel, Communication and Artificial Intelligence: Opportunities and Challenges for the 21st Century, in: *Communication +1*, 1 (1/2012), 1–26.

8   Alan Turing, On Computable Numbers, with an Application to the Entscheidungsproblem, in: *Proceedings Of The London Mathematical Society* 42 (2/1936), 230–265.

9   Alan Turing, Computing Machinery and Intelligence, in: *Mind* 59 (236/1950), 433–460.

10   Turing, Computing Machinery and Intelligence, 433.

11   Turing, Computing Machinery and Intelligence, 433.

becomes capable of successfully simulating a human being [...] in communicative exchanges with a human interrogator to such an extent that the interrogator cannot tell whether he is interacting with a machine or another human being, then that machine would, Turing concludes, need to be considered "intelligent."[12]

With his considerations, Turing pointed the way for AI research in the direction of communicative AI at an early stage, even if this perspective only became dominant in practice much later.[13]

However, although the idea to develop communicative artificial systems was not very prominent in the early days of AI, in the context of early approaches to automatic language processing, there were also initial attempts to develop dialog-capable systems. A famous precursor of today's communicative AI in this regard is the computer program ELIZA that Josef Weizenbaum developed in the 1960s. This particular program "simulated a psychotherapist's operation [by] returning the user's sentences in the interrogative form"[14] and was intended to investigate the limits and difficulties of natural language processing. However, although Weizenbaum wanted to "rob ELIZA [of] the aura of magic to which its application to psychological matter has to some extent contributed,"[15] many people who tested the program were fascinated by its output. Among these people were not only ordinary users of ELIZA but also information scientists who used the computer program and viewed its success as inspiration for the development of different kinds of early chatbots. Consequently, today, ELIZA is considered to be the world's first chatbot.[16] Nevertheless, the development of communicative machines led an entirely niche existence within the AI community for a long time. ELIZA's successors, such as PARRY and ALICE, still lacked sophisticated communication capabilities,[17] and for an extended period, ELIZA's limitations could not be overcome—that is, until the 1990s, when "transformations in computational infrastructure breathed new life into the project of designing humanlike, conversational artifacts."[18]

"Web-based and wireless technologies in particular inspired renewed attention to the interface as a site for novel forms of connection, both with and through com-

---

12   Gunkel, Communication and Artificial Intelligence, 5.

13   Guzman/Lewis, Artificial Intelligence and Communication, 71.

14   Eleni Adamopoulou/Lefteris Moussiades, Chatbots: History, technology, and applications, in: Machine Learning with Applications 2 (2020), see 2.

15   Joseph Weizenbaum, ELIZA--a computer program for the study of natural language communication between man and machine, in: Commun. ACM 9 (1/1966), 36–45, see 43.

16   Adamopoulou/Moussiades, Chatbots: History, technology, and applications.

17   Heung-yeung Shum/ Xiao-dong He/Li Di, From Eliza to Xiaolce: challenges and opportunities with social chatbots, in: Frontiers Inf Technol Electronic Eng 19 (1/2018), 10–26, see 12.

18   Lucy A. Suchman, Human-machine reconfigurations. Plans and situated actions (2nd edition), Cambridge 2007, see 206.

putational devices."[19] Accordingly, it was the birth of the world wide web that helped to transform the personal computer, which was designed for individual use, into a medium for communication. At the beginning, this was simply an "unintended byproduct of linking large computers to one another for security and information redundancy."[20] As Walther writes in an early paper about computer-mediated communication "operators found [that] they could send simple messages to one another,"[21] in addition to basic data transmission.

Very soon, this insight from the early days of the web led to the emergence of multiple forms of computer-mediated communication, including online computer games such as so-called multiuser dungeons (MUDs) that allowed users to engage in role-playing games online and gave the first online bots a home. As Sherry Turkle describes it in her famous book *Life on the Screen*, some MUD players left "behind small artificial intelligence programs called bots [...] running in the MUD that may serve as their alter egos, able to make small talk or answer simple questions."[22] The development of these early online bots that "perform[ed] roles previously reserved for people"[23] can probably be seen as the "birth hour" of web-based chatbots, embodied agents, smart speakers, and other digital devices that not only serve as communication mediums but also as communication partners.

Although the establishment of the world wide web was a starting point for the development of digital artificial communication partners, other technical advancements helped communicative AI to leave the digital space and enter the physical world. In particular, increased computer power and progress "made in programming as well as further technological advances in engineering"[24] led to new "possibilities of interfacing with people through sensors and actuators."[25] Consequently, two research fields that previously existed separately started to interweave with each other. One of these fields is communicative AI, and the other is robotics. For a long time, the latter was dedicated to industrial applications, mainly in the automotive

---

19   Suchman, Human-machine reconfigurations, 206.

20   Joseph B. Walther, Computer-Mediated Communication: Impersonal, Interpersonal, and Hyperpersonal Interaction, in: Communication Research 23 (1/1996), 3–43, see 5.

21   Walther, Computer-Mediated Communication: Impersonal, Interpersonal, and Hyperpersonal Interaction, 5.

22   Sherry Turkle, Life on the Screen. Identity in the Age of the Internet, New York 1995, see 12.

23   Turkle, Life on the Screen. Identity in the Age of the Internet, 88.

24   Michael Decker/Martin Fischer/Ingrid Ott, Service Robotics and Human Labor: A first technology assessment of substitution and cooperation, in: Robotics and Autonomous Systems 87 (2017), 348–354, see 348.

25   Nadia Magnenat-Thalmann, Social Robots: Their History and What They Can Do for Us, in: Hannes Werthner/Erich Prem/Edward A. Lee/Carlo Ghezzi (eds.), Perspectives on Digital Humanism, Cham 2022, 9–17, see 12.

industry, where robots substituted human labor[26] and took over standardized action sequences, such as spot welding.

Due to the aforementioned technical advancements, the situation looks very different today since robotic systems are now able to "take over non-standardized tasks previously reserved for humans."[27] Consequently, within the robotics community, "the focus has—at least partially—shifted from substitution to cooperation between human and machine."[28] In industrial contexts, such cooperation between humans and robots exists as a form of "co-work" that is not necessarily communicative.[29] Instead, so-called cobots are largely intended to help their human coworkers "with non-ergonomic, repetitive, uncomfortable or even dangerous operations."[30] For instance, they lift, move, or place workloads and thus support humans through reducing our physical effort or cognitive overload.[31]

However, robotic systems' new technical capacities have given rise to the idea that robots can be introduced to more complex work environments, focusing not only on (co-)operation but also on communication. Accordingly, today, robotic systems are not only found in factories but also in other domains, especially in the service sector, where robots are not only equipped with sensors and actuators but also with communicative capabilities. For instance, service robots can be used for entertainment purposes,[32] as assistants in the healthcare sector,[33] or as tour guides in museums or shopping centers.[34] Compared to industrial settings, the requirements for robots' capabilities in service domains are much higher.[35] This is due to the fact that in service contexts, the "tasks are often carried out in ever-changing environments (e.g. delivering luggage to a particular room), requiring navigational

---

26   Decker/Fischer/Ott, Service Robotics and Human Labor, 348.

27   Decker/Fischer/Ott, Service Robotics and Human Labor, 348.

28   Decker/Fischer/Ott, Service Robotics and Human Labor, 348.

29   Ingo Schulz-Schaeffer et al., The social construction of human-robot co-work by means of prototype work settings, TUTS – Working Papers 2 (2020), Berlin 2020, https://nbn-resolving.org/urn:nbn:de:0168-ssoar-71028-4 [last accessed: August 15, 2023].

30   Ales Vysocky/Petr Novak, Human Robot Collaboration in Industry, in: MM Science Journal (2016), 903–906, see 903.

31   Schulz-Schaeffer et al., The social construction of human-robot co-work by means of prototype work settings, 3.

32   Robert Bogue, The role of robots in entertainment, in: IR 49 (4/2022), 667–671.

33   Jane Holland et al., Service Robots in the Healthcare Sector, in: Robotics 10 (1/2021), 47.

34   Bogdan G. Draghici et al., Development of a Human Service Robot Application Using Pepper Robot as a Museum Guide, in: 2022 IEEE International Conference on Automation, Quality and Testing, Robotics (AQTR), Cluj-Napoca 2022, 1–5; Stefan Kopp et al., A Conversational Agent as Museum Guide – Design and Evaluation of a Real-World Application, in: Themis Panayiotopoulos et al. (eds.), Intelligent Virtual Agents, vol. 3661, Springer Berlin, Heidelberg 2005, 329–343.

35   Decker/Fischer/Ott, Service Robotics and Human Labor, 348.

capabilities for maneuvering through populated and sometimes constricted areas (e.g. [a] hotel elevator)."[36] Additionally, in service settings, robots often have to "interact with people to carry out their tasks (e.g. taking a food order or answering a question), requiring varying levels of capability and artificial intelligence."[37] This is exactly where robotics meets communicative AI and where robotic systems become "social" because they need to interact with humans in a humanlike way.[38]

3. From the Laboratory and into "the Wild"

Since the requirements for communicative AI in general and service robots in particular are very high, developers are facing various challenges related to issues such as robots' navigation of complex social environments[39] but also to issues like natural language processing[40] or the need for communicative AI systems to develop a "theory of mind" with regard to their interlocutors in order to understand their behavior and expectations.[41] Accordingly, communicative AI systems were unsurprisingly, for many years, "technologies-in-the-making" that mainly existed in laboratories within the research community. Ordinary people only encountered them as participants in controlled laboratory experiments. In the last couple of years, however, the first market-ready products were created, and companies started to sell robots, agent software, and digital assistants that are capable of performing real-world tasks and acting in real-world domains.

Hence, consumers nowadays find themselves chatting with smart assistants on company webpages or (more rarely) talking to robots in shopping malls or museums. Additionally, big tech firms have successfully introduced to the consumer market smart speakers that are capable of internet connectivity, can be controlled using spoken commands, and are able to connect with other devices. These systems have names such as Siri and Alexa and are promoted as communication partners.

36   Galen R. Collins, Improving human–robot interactions in hospitality settings, in: IHR 34 (1/2020), 61–79, see 62.

37   Collins, Improving human–robot interactions in hospitality settings, 62.

38   Cynthia L. Breazeal, Designing sociable robots, Cambridge 2002, see 2; Kerstin Dautenhahn, Methodology & Themes of Human-Robot Interaction: A Growing Research Field, in: International Journal of Advanced Robotic Systems 4 (1/2007a), 103.

39   Thibault Kruse et al., Human-aware robot navigation: A survey, in: Robotics and Autonomous Systems 61 (12/2013), 1726–1743.

40   Mary E. Foster, Natural language generation for social robotics: opportunities and challenges, in: Philosophical transactions of the Royal Society of London. Series B. Biological sciences 374 (1771/2019).

41   Cynthia L. Breazeal/Kerstin Dautenhahn/Takayuki Kanda, Social Robotics, In: Bruno Siciliano/Oussama Khatib (eds.), Springer Handbook of Robotics, vol. 16, Cham 2016, 1935–1972.

Up to the present, these systems have been sold millions of times and implemented within other systems, such as computers, smartphones, and tablets, demonstrating that communicative AI has now entered everyday life and that its domestication in private households and other social domains has already begun. Apparently, some users of these systems even develop feelings towards them and attribute personality to them.[42]

In this situation, many questions that have accompanied AI research from the very beginning are now being raised anew in an urgent manner. On a theoretical level, this not only concerns the question of the intelligence of machines, which has once again been prominently discussed in both popular and academic discourse in recent years. [43] Furthermore, the establishment of communicative AI in everyday settings also raises important ethical questions. With regard to bots and agents on the internet, for example, the extent to which it should be transparent to human users that they are dealing with an artificial counterpart is being discussed[44]. Linked to this, the extent to which it is possible and desirable for humans to enter into social relationships with machines is also up for debate. This includes dealing with emotional and erotic relationships. A critical look in this context is also taken at how relationships with machines differ from relationships between humans and what consequences the establishment of relationships with machines has for sociality and our human self-image.[45]

Such questions are not only of ethical relevance, as they point to the fact that communication and sociality are no longer only conceivable between human beings. This challenges basic theoretical assumptions of the communication and social sciences, which have so far focused primarily on communication and social relationships between humans and have based their theoretical conceptual apparatus on this.[46] In view of this, theorists have been discussing for several years how their own

---

42   Choi, Tae Rang/Minette E Drumwright, "OK, Google, why do I use you?" Motivations, post-consumption evaluations, and perceptions of voice AI assistants, in: *Telematics and Informatics* 62 (2021), 101628.

43   On google scholar, a search for "Turing Test" in February 2024 yields 15,500 hits for publications since 2020 alone.

44   Stefano Pedrazzi/Franziska Oehmer, Communication Rights for Social Bots?: Options for the Governance of Automated Computer-Generated Online Identities, in: *Journal of Information Policy* 10 (2020), 549–581.

45   Sherry Turkle, Alone together. Why we expect more from technology and less from each other, New York 2011; Cheok Adrian David/Karunanayaka Kasun/Zhang Emma Yann, Lovotics: Human-robot love and sex relationships, in: Patrick Lin/Keith Abney/R. Jenkins (eds.), *Robot Ethics 2.0: From Autonomous Cars to Artificial Intelligence*, Oxford 2017, 193–220; Blay Whitby, Do You Want a Robot Lover? The Ethics of Caring Technologies, in: Patrick Lin/Keith Abney (eds.), *Robot Ethics: MIT Press* 2012, 233–248.

46   Gesa Lindemann, The Analysis of the Borders of the Social World: A Challenge for Sociological Theory, in: *Journal for the Theory of Social Behaviour* 35 (1/2005), 69–98; Andreas Hepp, Artificial

theoretical concepts need to be adapted in order to be able to adequately take into account the (communicative) relationships between humans and machines. How such adaptations should look like is controversial and the subject of ongoing debate in the individual disciplines.[47]

However, challenges arise not only for theoretical reasoning, but also for empirical research when communicative AI takes its path 'into the wild'. This is because the conditions of research change dramatically when AI systems enter everyday contexts. Previously, research took place primarily under controlled conditions in the laboratories of the technical sciences. Accordingly, experimental methods were (and still are) the standard, when it comes to analyzing human-machine communication.[48] Although it appeared suitable to mainly rely on experimental research methods under these circumstances, the investigation of communicative AI in everyday life demands other methods.[49] Human–machine encounters in authentic everyday life settings can hardly be investigated based on methods that rely on controlled laboratory conditions.[50] Accordingly, the new situation of communicative AI, given its entrance into everyday life, calls for new, "less constrained, open-ended and more exploratory studies"[51] as compared to the methods traditionally applied in the field. Respective studies need to be able to investigate not only artificial and restricted interaction scenarios under laboratory conditions, but naturally occurring interactions in real-world settings "to shed light on the situated nature of human-robot/agent interaction and the participants' communicative conduct and their micro-practices of interactional coordination".[52]

---

companions, social bots and work bots: communicative robots as research objects of media and communication studies, in: *Media Culture Society* 42 (7–8/2020), 1410–1426.

47   Hepp, Artificial companions, social bots and work bots, 1410–1426; Guzman/Lewis, Artificial Intelligence and Communication, 70–86; Florian Muhle, Sozialität von und mit Robotern? Drei soziologische Antworten und eine kommunikationstheoretische Alternative, in: *Zeitschrift für Soziologie* 47 (3/2018), 147–163; Michaela Pfadenhauer, On the Sociality of Social Robots. A Sociology-of-Knowledge Perspective, in: *Science, Technology & Innovation Studies* 10 (1/2014).

48   Christoph Bartneck/Tony Belpaeme/Friederike Eyssel/Takayuki Kanda/Merel Keijsers/Selma Šabanović, Human-robot interaction. An Introduction, Cambridge 2020, see 127.

49   Malte Jung/Pamela Hinds, Robots in the Wild: A Time for More Robust Theories of Human-Robot Interaction, in: ACM Trans. Hum.-Robot Interact. 7 (1/2018), Article 2; Karola Pitsch, Interacting with Robots and Virtual Agents? Robotic Systems in Situated Action and Social Encounters, in: Mensch und Computer 2019—Workshopband, Bonn 2019, 341–342.

50   Jung/Hinds, Robots in the Wild: A Time for More Robust Theories of Human-Robot Interaction.

51   Kerstin Dautenhahn, Robots in the Wild. Exploring Human-Robot Interaction in Naturalistic Environments, in: IS 10 (3/2009), 269–273, see 270.

52   Pitsch, Interacting with Robots and Virtual Agents, 342.

4. Approach and Structure of the Book

In view of the theoretical and methodological challenges described above, which arise with the establishment of communicative AI "in the wild", this anthology focuses on the methodological challenges. In line with other scholars in the field, we assume that particularly qualitative social scientific methods developed explicitly for investigating naturally occurring interactions[53] appear to be suitable for investigating communicative AI in (inter)action. How studies can look and the kinds of insights such research can provide constitute the key subject of this anthology, which brings together contributions from a still small but growing community of researchers who are committed to exploring communicative AI "in action" by using and adapting qualitative (and mostly ethnographic) methods.

The focus on methodological questions and empirical case studies does not argue against the relevance of the theoretical and ethical reasoning in the context of communicative AI. Rather, we assume that the discussion of methodological questions and the empirical examination of human-machine communication can make a significant contribution to informing the theoretical and ethical discourse. For example, empirical analyses of human-machine communication can shed light on the extent to which this continues to differ from interpersonal communication and thus also enrich debates about the intelligence of machines in a specific way with detailed findings. Similarly, ethnographic observations of the use of communicative AI in elderly care facilities can provide indications of ethical implications that could not be achieved through theoretical considerations alone.

Originally, the contributions of this book were intended to be presentations given at a conference to be held in 2020, as the final conference of the Deutsche Forschungsgemeinschaft (German Research Foundation) funded research project Communication at the Borders of the Social World. Unfortunately, due to the COVID-19 pandemic, the conference could not take place. Instead, during times of "social distancing," the idea for this anthology was born, and most of the colleagues who we originally wanted to meet and get to know in person during the conference decided to participate in this project. However, it took over two years of writing and editing (still under the conditions induced by the global health crisis) before the publication was camera-ready. We are convinced that the result is worth the time and work everyone involved has invested in the project, and we would like to thank everybody for their efforts and patience. We hope that this book will contribute to making social science approaches that utilize qualitative methods more visible and established in the context of communicative AI research. In our opinion, the articles collected in this anthology impressively show that such methods are

---

53   Manja Lohse et al., Gerhard, Improving HRI design by applying Systemic Interaction Analysis (SInA), in: IS 10 (3/2009), 298–323; Pitsch, Interacting with Robots and Virtual Agents?.

desirable and that they contribute to new and deeper insights into the specifics of the contemporary forms and problems of human–machine communication (see the concluding chapter). In essence, the contributions address (with different emphases) the following key questions:

- What are adequate methods for investigating communicative AI in (inter-)action?
- What are forms and characteristics of interaction with communicative AI?
- How are encounters with communicative AI framed and shaped by institutional settings?
- How can interaction with communicative AI in different settings be compared?

These questions are answered on the basis of eight articles written by international scholars who deal empirically and conceptually with different variants of communicative AI. Most of the contributions present empirical case studies that deal with one particular technological system in (inter-)action. These contributions are accompanied by two methodological articles with a broader focus. Accordingly, the anthology is divided into four parts. Parts I–III gather contributions to particular communicative AI technologies, namely social robots, embodied conversational agents, and smart speakers. Part IV contains two contributions that are not based on single case studies but rather focus primarily on methodological issues pertinent to analyzing human–machine communication. The anthology is concluded by a short article by Florian Muhle and Indra Bock, in which the authors summarize the main insights of the book.

Part I engages with social robots in (inter-)action. In Chapter 1, "Programming Engagement: Shaping Human–Robot–Public Interaction in a Smart City Robot Competition," Carlos Cuevas-Garcia and Cian O'Donovan deal with human–robot–public interaction. Based on a situational analysis of SciRoc, a smart city robotics competition organized by the European Robotics League in partnership with a city council in the United Kingdom and a number of academic and commercial sponsors, they identify three modes of human–robot–public engagement: embracing engagement, bypassing engagement, and prefiguring engagement. The authors show that and how these three modes of engagement in turn revealed and were shaped by different logics of social ordering, namely conviviality, control, and care. In this sense, they impressively show how the competition's organizers predetermined the possibilities of human–robot interaction.

In Chapter 2, "Towards Placing Service Robots in Elderly Care Facilities," Rosalyn M. Langedijk and Kerstin Fischer present three case studies about the development of service robots for use in elderly care facilities and their actual employment there. Using an ethnographic approach, they shed light on real users' needs and show how difficult it is to implement robotic solutions that support real-world tasks and fit the everyday needs of elderly care facilities. Based on these insights, the authors suggest recommendations for future real-world testing. Additionally, they share their reflections on   ethical issues and preparations regarding their field trials and hence provide important information for researchers aiming to enter the field of human–robot interaction.

Part II is dedicated to the empirical analysis of embodied agents in (inter-)action. In his chapter, "Mixed Methods for Mixed Realities: The Analysis of Multimodal Interactions With Embodied Conversational Agents," Jonathan Harth deals with multimodal interactions with anthropomorphic virtual agents. Whereas existing research paradigms for the analysis of human–agent interaction mainly focus only on the user's perception of interaction, he presents a methodological approach that focuses on the emergent interaction processes themselves and allows for analysis of both the relationship level as well as the content level in human–agent interaction. As the author argues, this approach renders it possible to identify potential discrepancies between the user's individual experiences and the physically expressed behavior during interactions, which might also help improve agent systems' communicative capabilities.

Florian Muhle, Indra Bock, and Henning Mayer are also interested in the analysis of interactions between embodied agents and humans. However, in their chapter, "Investigating the Architecture-for-Interaction of an Embodied Conversational Agent," they propose a slightly different approach than that espoused in Jonathan Harth's contribution. Based on the observation that comprehension problems are normal in human–machine interaction and that technical systems' usability often fails to live up to users' expectations, the authors present an approach for analyzing communicative AI systems' architecture-for-interaction, which can be used to show in detail at which point and why communicative problems in human–machine encounters arise. They exemplify this by means of a case study in which they examine the beginning of an encounter between a visitor and an artificial museum guide in a computer museum.

The next two chapters are concerned with smart speakers. On the one hand, these human–machine interfaces are less humanoid than social robots and embodied agents, but on the other hand, they are very successful and market-ready products that have already made their way into millions of households and are thus presently the most established form of communicative AI. However, although smart speakers are commercialized and often treated as conversational interfaces, Brian L. Due and Louise Lüchow show in their chapter, "VUI-Speak: There Is Nothing Conversational About 'Conversational User Interfaces,'" that various devices' voice-based operation shows features that are quite different from everyday conversation between humans. More precisely, they apply video ethnographic studies and ethnomethodological conversation analysis of blind people's natural use of Google Home to investigate the exchange between humans and machines in a fine-grained manner. On this basis, they identify a phenomenon, which they describe as "VUI-speak," through which people accommodate devices. That is, it is not the "smart" machine that adapts to users and their needs. Instead, it is the other way around, with intelligent users adapting to the machine and its constraints to operate the device successfully.

Miriam Lind considers smart speakers in a slightly different methodological manner. In her chapter, "Doing Family on Unfamiliar Terrain: The Constitution and Contestation of Kinship Among Two Humans, Two Cats, and a Voice Assistant", she presents an autoethnographic pilot study on the doing and undoing of family and kinship between humans, cats, and Amazon's Alexa in a private household. Based on the logs that Amazon's Alexa program automatically stores as well as "reflexive investigation", the author analyzes the interaction and communicative behavior in a household and asks in which ways the artificial companion is included and excluded in practices of doing family, how technical obstacles and communication breakdowns affect these practices, and how human–machine interaction is embedded in human beliefs and attitudes towards family, technology, and interaction. In doing so, she provides a critical approach to human–machine interaction "in the wild" and examines how the introduction of voice assistants into the privacy of homes and into family systems impacts our understanding of communication and the "fragile institutionalization" of family.

As mentioned above, Part IV of the anthology is dedicated to broader methodological issues. Whereas the chapters in the first three parts primarily focus on case studies, the chapters in this part share a broader focus. Arne Maibaum, Philipp Graf, and René Tuma deal with the use of video recording in the research field of human–robot interaction (or HRI) as the title of their chapter, "On the Use of Videography in HRI," suggests. The starting point of their argument is that video recording is common and widespread in the field of human–robot interaction, but at the same time, it is used for very different purposes and not in a systematical manner. Against this background, the authors argue that a methodological reflected use of videos, combined with an ethnographic research design, is necessary to realize the full potential of video data collection and interpretation because this is the only way to realize more accurate evaluations and explorations of human–robot interaction situations. Especially now that robots are entering new real-world institutional contexts, such an approach appears to be necessary. Drawing on examples from their research, the authors demonstrate the plausibility of their considerations and elaborate on the importance of ethnography for videographic work in HRI to interpret and make sense of the recorded data, as well as for the conception of video recordings.

Dafna Burema considers the question of using secondary data in the study of human–machine interaction. In her chapter, "Studying Interaction Indirectly: The Relevance of Secondary Data for Studying Human–Robot Interaction Empirically," she argues for using secondary data when empirically studying human–robot interaction. Her first argument for the use of secondary data is restricted access to the field. For researchers who lack financial or symbolic capital in particular, it is difficult to study human–robot interaction using primary data, and access to existing data would make it easier to enter the research field. Her second argument for using secondary data is that such data allow for comparisons and would thus yield more generalizable insights. Although human–robot interaction research is first and foremost based on case studies—as the chapters in the first three parts of this anthology show—the use of secondary data could help to broaden the analytical focus and obtain a bigger picture of issues that are *typical, systemic,* or *recurring* in HRI.

Finally, Florian Muhle and Indra Bock summarize the insights provided by the chapters of this anthology. They highlight both the strengths and weaknesses of ethnographically oriented research in the field of human-machine communication. In addition, they make clear that communication with communicative AI still differs significantly from human interaction, which shows the (technical) challenges that need to be overcome in order to establish communicative AI in private and institutional contexts outside the laboratory.

Bibliography

Adamopoulou, Eleni/Moussiades, Lefteris, Chatbots: History, technology, and applications, in: Machine Learning with Applications 2 (2020), 100006.

Afonso, Nuno/Prada, Rui, Agents That Relate: Improving the Social Believability of Non-Player Characters in Role-Playing Games, in: Scott M. Stevens/Shirley J. Saldamarco (eds.), *Entertainment Computing—ICEC 2008*, vol. 5309, Berlin, Heidelberg 2009, 34–45.

Bogue, Robert, The role of robots in entertainment, in: IR 49 (4/2022), 667–671.

Breazeal, Cynthia L., *Designing sociable robots*, Cambridge 2002.

Breazeal, Cynthia/Dautenhahn, Kerstin/Kanda, Takayuki, Social Robotics, In: Bruno Siciliano/Oussama Khatib (eds.), *Springer Handbook of Robotics*, vol. 16, Cham 2016, 1935–1972.

Cheok, Adrian David/Karunanayaka, Kasun/Zhang, Emma Yann, Lovotics: Human-robot love and sex relationships, in: Patrick Lin/Keith Abney/R. Jenkins (eds.), *Robot Ethics 2.0: From Autonomous Cars to Artificial Intelligence*, Oxford 2017, 193–220.

Choi, Tae Rang/Drumwright Minette E, "OK, Google, why do I use you?" Motivations, post-consumption evaluations, and perceptions of voice AI assistants, in: *Telematics and Informatics* 62 (2021), 101628.

Collins, Galen R., Improving human–robot interactions in hospitality settings, in: *IHR* 34 (1/2020), 61–79.

Dautenhahn, Kerstin, Methodology & Themes of Human-Robot Interaction: A Growing Research Field, in: *International Journal of Advanced Robotic Systems* 4 (1/2007a).

Dautenhahn, Kerstin, Robots in the Wild. Exploring Human-Robot Interaction in Naturalistic Environments, in: *IS* 10 (3/2009), 269–273.

Dautenhahn, Kerstin, Socially intelligent robots: dimensions of human–robot interaction, in: *Philosophical Transactions of the Royal Society B. Biological Sciences* 362 (1480/2007b), 679–704.

Decker, Michael/Fischer, Martin/Ott, Ingrid, Service Robotics and Human Labor: A first technology assessment of substitution and cooperation, in: *Robotics and Autonomous Systems* 87 (2017), 348–354.

Draghici, Bogdan G./Dobre, Alexandra E./Misaros, Marius/Stan, Ovidiu P., Development of a Human Service Robot Application Using Pepper Robot as a Museum Guide, in: *2022 IEEE International Conference on Automation, Quality and Testing, Robotics (AQTR)*, Cluj-Napoca 2022, 1–5.

Foster, Mary Ellen, Natural language generation for social robotics: opportunities and challenges, in: *Philosophical transactions of the Royal Society of London. Series B. Biological sciences* 374 (1771/2019).

Gunkel, David J., Communication and Artificial Intelligence: Opportunities and Challenges for the 21st Century, in: *Communication +1*, 1 (1/2012). 1–26.

Guzman, Andrea L./Lewis, Seth C., Artificial Intelligence and Communication: A Human–Machine Communication Research Agenda, in: *New Media & Society* 22 (1/2020), 70–86.

Hepp, Andreas, Artificial companions, social bots and work bots: communicative robots as research objects of media and communication studies, in: *Media Culture Society* 42 (7–8/2020), 1410–1426.

Holland, Jane/Kingston, Liz/McCarthy, Conor/Armstrong, Eddie/O'Dwyer, Peter/Merz, Fionn/McConnell, Mark, Service Robots in the Healthcare Sector, in: *Robotics* 10 (1/2021), 47.

Jung, Malte/Hinds, Pamela, Robots in the Wild: A Time for More Robust Theories of Human-Robot Interaction, in: *ACM Trans. Hum.-Robot Interact.* 7 (1/2018), Article 2.

Kempt, Hendrik, *Chatbots and the Domestication of AI. A Relational Approach*, Cham 2020.

Kopp, Stefan/Gesellensetter, Lars/Krämer, Nicole C./Wachsmuth, Ipke, A Conversational Agent as Museum Guide – Design and Evaluation of a Real-World Application, in: *Themis Panayiotopoulos/Jonathan Gratch/Ruth Aylett/Daniel Ballin/Patrick Olivier/Thomas Rist* (eds.), *Intelligent Virtual Agents*, vol. 3661, Springer Berlin, Heidelberg 2005, 329–343.

Kruse, Thibault/Pandey, Amit Kumar/Alami, Rachid/Kirsch, Alexandra, Human-aware robot navigation: A survey, in: *Robotics and Autonomous Systems* 61 (12/2013), 1726–1743.

Lindemann Gesa, The Analysis of the Borders of the Social World: A Challenge for Sociological Theory, in: *Journal for the Theory of Social Behaviour* 35 (1/2005), 69–98.

Lohse, Manja/Hanheide, Marc/Pitsch, Karola/Rohlfing, Katharina J./Sagerer, Gerhard, Improving HRI design by applying Systemic Interaction Analysis (SInA), in: *IS* 10 (3/2009), 298–323.

Magnenat-Thalmann, Nadia, Social Robots: Their History and What They Can Do for Us, in: Hannes Werthner/Erich Prem/Edward A. Lee/Carlo Ghezzi (eds.), *Perspectives on Digital Humanism*, Cham 2022, 9–17.

McCorduck, Pamela, *Machines Who Think. A Personal Inquiry into the History and Prospects of Artificial Intelligence*, Natick 2004.

Muhle Florian, Sozialität von und mit Robotern? Drei soziologische Antworten und eine kommunikationstheoretische Alternative, in: *Zeitschrift für Soziologie* 47 (3/2018), 147–163.

Muhle, Florian/Bock, Indra, Intuitive Interfaces? Interface Design and its Impact on Human-Robot Interaction, in: *Mensch und Computer 2019 – Workshopband*, Bonn 2019, 346–347.

Pedrazzi, Stefano /Oehmer, Franziska, Communication Rights for Social Bots?: Options for the Governance of Automated Computer-Generated Online Identities, in: *Journal of Information Policy* 10 (2020), 549–581.

Pfadenhauer, Michaela, On the Sociality of Social Robots. A Sociology-of-Knowledge Perspective, in: *Science, Technology & Innovation Studies* 10 (1/2014).

Pitsch, Karola, Interacting with Robots and Virtual Agents? Robotic Systems in Situated Action and Social Encounters, in: *Mensch und Computer 2019—Workshopband*, Bonn 2019, 341–342.

Prada, Rui/Paiva, Ana, Human-Agent Interaction: Challenges for Bringing Humans and Agents Together, in: *HAIDM—3rd International Workshop on Human-Agent Interaction Design and Models held at AAMAS'2014—13th International Conference on Autonomous Agents and Multi-Agent Systems*, Paris 2014.

Schulz-Schaeffer, Ingo/Meister, Martin/Wiggert, Kevin/Clausnitzer, Tim, The social construction of human-robot co-work by means of prototype work settings, *TUTS – Working Papers* 2 (2020), Berlin 2020, https://nbn-resolving.org/urn:nbn:de:0168-ssoar-71028-4 [last accessed: August 15, 2023].

Shum, Heung-yeung/He, Xiao-dong/Di Li, From Eliza to XiaoIce: challenges and opportunities with social chatbots, in: *Frontiers Inf Technol Electronic Eng* 19 (1/2018), 10–26.

Suchman, Lucy A., *Human-machine reconfigurations. Plans and situated actions* (2nd edition), Cambridge 2007.

Turing, Alan, Computing Machinery and Intelligence, in: *Mind* 59 (236/1950), 433–460.

Turing, Alan, On Computable Numbers, with an Application to the Entscheidungsproblem, in: *Proceedings Of The London Mathematical Society* 42 (2/1936), 230–265.

Turkle, Sherry, *Alone together. Why we expect more from technology and less from each other*, New York 2011.

Turkle, Sherry, *Life on the Screen. Identity in the Age of the Internet*, New York 1995.

Vysocky, Ales/Novak, Petr, Human Robot Collaboration in Industry, in: *MM Science Journal* (2016), 903–906.

Wachsmuth, Ipke, Embodied Cooperative Systems: From Tool to Partnership, in: Catrin Misselhorn (ed.), *Collective Agency and Cooperation in Natural and Artificial Systems. Explanation, Implementation and Simulation*, Cham, Heidelberg, New York, Dordrecht, London 2015, 63–79.

Walther, Joseph B., Computer-Mediated Communication: Impersonal, Interpersonal, and Hyperpersonal Interaction, in: *Communication Research* 23 (1/1996), 3–43.

Weizenbaum, Joseph, ELIZA--a computer program for the study of natural language communication between man and machine, in: *Commun. ACM* 9 (1/1966), 36–45.

Whitby, Blay, Do You Want a Robot Lover? The Ethics of Caring Technologies, in: Patrick Lin/Keith Abney (eds.), *Robot Ethics*: MIT Press 2012, 233–248.

Zhao, Shanyang, Humanoid social robots as a medium of communication, in: *New Media & Society* 8 (2006), 401–419.

Part I: Social Robots in (Inter-)Action

Programming Engagement: Shaping Human-Robot-Public Interaction in a Smart City Robot Competition

Carlos Cuevas-Garcia, Cian O'Donovan

Abstract *This chapter presents a situational analysis of SciRoc, the first ever "**Smart city Ro**bots competition", organized by the European Robotics League (ERL) in partnership with Milton Keynes City Council in the United Kingdom and a number of academic and commercial sponsors. Besides this competition, we use data collected during other ERL competitions in test beds and living labs in Madrid, Oldenburg, Bristol, mainstream media reporting and extensive conversations with participants. We argue that since competitions are constituted by different sets of rules, and since these rules intersect with the values, practices, assumptions, politics, and interests of their sponsors and organizers, they are appropriate sites for studying the institutional shaping of human-robot-public interaction. We identified three modes of human-robot-public engagement: embracing engagement, an open and attentive form of engagement that was sensitive to the needs, interests, and concerns of various participants, sponsors, and members of the audience. Second, bypassing engagement, a more constrained and constraining form of engagement that limited the possibilities of mutual understanding between competition participants and the various publics. Third, prefiguring engagement, a variety of previous commitments and expectations that brought the event into being and gave it shape, but that rigidly framed the ways in which publics and participants could engage with each other. These three modes of engagement in turn revealed and were shaped by different logics of social ordering, namely conviviality, control, and care.*

1. Introduction

On 18 September 2019, 18 year-old hairdressing student Leila Ahmed walked out from *The Hair and Makeup Academy* and into *MK:Centre* for her daily wander in this shopping mall, the largest in Milton Keynes and a central place for social life in this 1960s-founded English town. Heading for the *Costa* café, Leila was intent on a strawberry iced infusion and a browse through some shops before making her way home.

As Leila approached *Costa*, she encountered what for sure was one of the most bizarre set ups she ever saw in the shopping mall: The entire exhibition hall, an area of nearly 60x40 m, was taken over by a large gated setting composed of a portable il-lumination structure reminiscent of a music concert stage, an interior wall formed by 1.20 m high hoardings, and a number of massive posters walls of nearly 5x6-10 m that sectioned the area. On the posters were cartoon-like illustrations of peo-ple enjoying beverages in a café (see Fig. 1), taking the lift (see Fig. 3), and receiving paramedic attention (see Fig. 4). On each, the poster's larger-than-life inhabitants interacted harmoniously with docile, friendly looking, big-eyed robots.

Other small hoardings with prints from *Costa* and the European Union hanged inside these premises, and one section of this odd setup was fully covered by a 5 m high protection net cage (see Fig. 4). The whole area was surrounded by barrier belts that read "for your own safety please do not enter". In the middle of all this paraphernalia, at least 60 people ran moving around unused furniture, flat screens, and spare lamps. Some others installed workplaces with laptops, and some of them unboxed and assembled, what seemed to be, real-life robots (see Figs. 1–4).

Fig. 1: The SciRoc competition arena from different angles (the authors, 2019)

Fig. 2: The SciRoc competition arena from different angles (the authors, 2019)

Fig. 3: The SciRoc competition arena from different angles (the authors, 2019)

Fig. 4: The SciRoc competition arena from different angles (the authors, 2019)

Today was the first of several days in which Leila's daily routine would be disrupted by a complex and multi-layered event, to which she came back on several occasions to observe and to make sense, little by little, of what it had to offer. Before enrolling in *the Academy*, she had considered studying computer science, and the event's visual design reminded her of a number of recurring themes from her time investigating computer science pamphlets and degree program webpages: the digital transformation is around the corner; social robots and artificial intelligence will make tremendous impacts in our daily lives; a young workforce must be prepared to take part in this exciting, if scary, technological revolution. Leila was also concerned about the idea of having robots everywhere – she was aware that hackers could take control of them and data wasn't always kept private. But most of all, Leila was intrigued by the images of people living, being and working with robots in what looked to be the most convivial of smart cities[1].

---

1   We thank the editors of this volume for raising up the issue that hairdressing is a traditionally gendered role. We decided to include this detail because it was indeed hairdressing that a young member of the audience – who here we call Leila – studied. Yet, she also had considered studying computer science. We include this detail to highlight that ordinary members of the audience have rich and fascinating lives, they are intelligent and critical, and scientific and technological higher education might be in their interests even though they are enrolled in non-scientific occupations.

The event that Leila walked into at the MK:Centre was the product of a multi-institutional coordination effort: *SciRoc*, the first ever "**S**mart **c**ity **Ro**bots **c**ompetition", organized by the *European Robotics League (ERL)* in partnership with Milton Keynes City Council and a number of academic and commercial sponsors.[2] The *SciRoc* website stated that "recent developments, such as autonomous cars and service robots, provide […] evidence that smart cities are indeed a privileged environment for the introduction of robotic technologies"[3]. *SciRoc* intended to advance "the integration of autonomous systems in smart cities" by examining "difficulties of dealing with complex and large scale scenarios"[4]. The objective was to assess "how robots can integrate and co-operate with a complex city environment [and] how robots can act both as data collectors and data consumers of the cities' digital hubs"[5]. Most importantly, *SciRoc* aimed to present "the first robotics challenges where robots will interact with ordinary people (i.e. customers of the shopping mall) […] offering unique opportunities to boost the robots' social acceptance (as companions or helpers) and the smart interaction with other devices and resources"[6].

But this agenda seemed to arrive to the MK:Centre with important issues already decided. First, that accelerating the social acceptance of robots and smart cities was something desirable. Second, that competition was an appropriate means by which to do this. Third, that reproducing a convivial smart city was a form of engaging the public in which people could give their views and make decisions about technological transformations affecting their day to day lives (see Sclove[7] for a more detailed treatment of this issue).

We wanted to know how this event could address the concerns of a young citizen like Leila. More critically, following a trend on experimental involvement of society in science and innovation[8], we were there to ask *how is public engagement enacted*

---

2   Matthew E. Studley/ Hannah Little, Robots in Smart Cities, in: Maria I. Aldinhas Ferreira (ed.), *How Smart Is Your City? Technological Innovation, Ethics and Inclusiveness*, Cham 2021, 75–88.

3   Damian Dadswell, ERL Smart Cities (2018a), URL: http://instituteofcoding.open.ac.uk/ [last accessed: August 15, 2023].

4   Dadswell, ERL Smart Cities.

5   Dadswell, ERL Smart Cities.

6   Dadswell, ERL Smart Cities.

7   Richard E. Sclove, *Democracy and Technology* (1st edition), New York 1995.

8   Franziska Engels/Alexander Wentland/Sebastian M. Pfotenhauer, Testing Future Societies? Developing a Framework for Test Beds and Living Labs as Instruments of Innovation Governance, in: *Research Policy* 48 (9/2019), 1–25; Brice Laurent et al., The Test Bed Island: Tech Business Experimentalism and Exception in Singapore, in: *Science as Culture* 30 (3/2021), 367–90; Harriet Bulkeley/ Vanesa Castán Broto, Government by Experiment? Global Cities and the Governing of Climate Change: Government by Experiment?, in: *Transactions of the Institute of British Geographers* 38 (3/2013), 361–75; Aidan H. While/Simon Marvin/Mateja Kovacic, Urban Robotic Experimentation: San Francisco, Tokyo and Dubai, in: *Urban Studies* 58 (4/2021), 769–86.

during robot competitions? And how do robot competitions contribute to configuring the social order in the smart city?

These are questions about how practices of public engagement with, and public understanding of science[9] intersect with ideas of how societies and technologies are and should be ordered. These guide our inquiry over the remainder of this chapter. Our departure point is this: since competitions are constituted by different sets of rules, some explicit and some that go entirely unsaid, and since these rules intersect with the values, practices, assumptions, politics, and interests of their sponsors and organizers, competitions are appropriate sites for studying the institutional shaping of human-robot-public interaction.

2. Research Design and Methodology: A Situational Analysis of Human-Robot-Public Engagement

We proceed in this study by way of reporting a case narrative drawn from a situational analysis of this public robot competition. Briefly, situational analysis is an interpretive, grounded theory approach that offers a materialist constructionism by mapping the social and material phenomena that make a difference in a given situation.[10] A situation is an inventory of communities and activities that happen in a space that is considered relationally as shaped through shared discourse.

The analysis proceeds via a series of mapping techniques. First, situation maps of major human, non-human, discursive elements. Second at a meso-level collective actors and their shared or contended commitments are mapped. A third layer of mapping follows locating the major positions taken by actors in the data, noting concerns and controversies in the situation.

The analytic goal in this study was to specify which entities – of varying scale and composition – make a difference to the situation of the robot competition from the perspective of the people involved. In this case the situation consists of the settings, venues, devices, scenarios, and narratives performed by actors as they framed and enacted the competition. We pay close attention to how they articulated the ways in which robots interacted with their programmers and with diverse publics.

The method was chosen as it lets us go beyond the usual suspects of highly bounded sociological framings of organizations, institutions and social movements and allows us think about discourse based social action. Through situational analy-

---

9   Sarah R. Davies, An Empirical and Conceptual Note on Science Communication's Role in Society, in: *Science Communication* 43 (1/2021), 116–33.

10   Adele E. Clarke, *Situational Analysis: Grounded Theory After the Postmodern Turn* (1st edition), Thousand Oaks 2005.

sis we understand situations as distributed action and accomplishments, which are produced through the coming together of heterogeneous elements.

This is helpful as we consider robotics for seemingly different social purposes and investigate the social and material conditions that make certain practices and routines acceptable at any given time – what we call below logics of social ordering. The narrative teases out how the participants and organizers of the competition brought together their preferences, motivations and expectations; where these came from, and what tensions they brought into the situation.

In other words, using situational analysis we examine how this competition was brought into being and how it shaped, and was shaped by modes of human, robot and public engagement in particular ways – cleaving to or against certain ideas and expectations of how people and artifacts should be configured through practices, performances and standardizations aiming to establish social orders.[11] Throughout this text we explain the different modes of engagement in relation to how they weave together three different logics of social ordering: *conviviality* (appreciating mutualistic autonomy), *care* (where connections are prioritized over hierarchies), and *control* (as an imperative to maintain fictitious borders and hierarchies between subjects and objects).[12] We return to these logics in the discussion.

The data used for the situational analysis came from field notes and interviews we made at ERL competition sites in Madrid, Oldenburg, Bristol, and Milton Keynes as well as academic literature about robotics competitions, mainstream media reporting and extensive conversations with participants as part of a broader project investigating innovation practices and policies in robotics at locations across Europe from 2018–2021.[13]

This multi-method approach enabled us to open up the competition at the MK:Centre to examine its different components, to get a better grasp of the organizers' roles, and what it means to make a public engagement event under the wider frame of the smart cities discourse, in Milton Keynes in particular. We proceed by accounting for how exactly smart cities and robot competitions have ended up in a shopping mall in Milton Keynes.

11   Sheila Jasanoff, *States of Knowledge: The Co-Production of Science and the Social Order*, London, New York 2004; Lucy Suchman, *Human-Machine Reconfigurations: Plans and Situated Actions* (2nd edition), Cambridge, New York 2006.

12   Saurabh Arora et al., Control, Care, and Conviviality in the Politics of Technology for Sustainability, in: *Sustainability: Science, Practice, and Policy* 16 (1/2020), 247–62.

13   Ola Michalec/Mehdi Sobhani/Cian O'Donovan, What Is Robotics Made of? The Politics of Interdisciplinary Robotics Research, in: *Humanities & Social Sciences Communications* 8 (2021), article 65; Cian O'Donovan, Accountability and Neglect in UK Social Care Innovation, in: *Policy Press* 7 (1/2022), 67–90; Carlos Cuevas-Garcia/Federica Pepponi/Sebastian M. Pfotenhauer, Maintaining Innovation: How to Make Sewer Robots and Innovation Policy Work in Barcelona, in: *Social Studies of Science* 54(3):352–376.

3. Smart Cities and Robot Competitions

The *SciRoc* competition is worthwhile exploring because it represents a particular instance in which robot competitions were brought into the larger techno-political vision of the smart city. Most importantly, in this competition human-robot interaction was at the Centre of the smart cities vision. Recent literature has looked at the role of robots in cities.[14] Yet, since public engagement events bringing together these two elements are uncommon, there is scant literature on the topic. This is where our study makes a novel contribution.

The notion of smart cities refers to urban environments in which digital technologies and infrastructures make all sorts of transport, energy, and communication services more efficient. Cities around the world have implemented different measures to become "smarter", attracting new companies and investors to take part in these transformations.[15] Some smart city initiatives are led by the private sector, others by the government, and others rely strongly on citizen initiatives.[16] While novel technologies can indeed improve citizens' quality of life, they also imply more permanent surveillance and the delegation of public services to foreign private companies, pointing to a number of dystopian scenarios.[17] Studies that observe that the dominant smart city imaginary was produced by IBM and Cisco call for larger citizen participation in the production of "counter-narratives that open up space for alternative values, designs, and models"[18]. Yet, citizen participation in smart city

---

14   Rachel Macrorie/Simon Marvin/Aidan While, Robotics and Automation in the City: A Research Agenda, in: *Urban Geography* 42 (2/2019), 1–21; While/Marvin/Kovacic, Urban Robotic Experimentation; Simon Marvin et al., *Urban Robotics and Automation: Critical Challenges, International Experiments and Transferable Lessons for the UK*, EPSRC UK Robotics and Autonomous Systems (RAS) Network.

15   Vincent Mosco, *Smart City in a Digital World*, Bingley 2019; Adrian Smith/Pedro P. Martín, Going Beyond the Smart City? Implementing Technopolitical Platforms for Urban Democracy in Madrid and Barcelona, in: *Journal of Urban Technology* (2020), 1–20; Burcu Baykurt/ Christoph Raetzsch, What Smartness Does in the Smart City: From Visions to Policy, in: *Convergence* 26 (4/2020), 775–89.

16   Adrian Smith, Smart Cities Need Thick Data, Not Big Data, in: *The Guardian*, 18.04.2018, URL: http://www.theguardian.com/science/political-science/2018/apr/18/smart-cities-need-thick-data-not-big-data; Mosco, Smart Cities in a Digital World.

17   Mosco, Smart Cities in a Digital World; Britt Paris, The Internet of Futures Past: Values Trajectories of Networking Protocol Projects, in: *Science, Technology, & Human Values* 46 (5/2020), 1021–1047; Robert Muggah/Greg Walton, Smart' Cities Are Surveilled Cities, URL: https://foreignpolicy.com/2021/04/17/smart-cities-surveillance-privacy-digital-threats -internet-of-things-5g/ [last accessed: August 15, 2023].

18   Jathan Sadowski/ Roy Bendor, Selling Smartness: Corporate Narratives and the Smart City as a Sociotechnical Imaginary, in: *Science, Technology, & Human Values* 44 (3/2019), 540–63, see 540.

politics can take different forms[19] and it occurs under conditions of "unequal relations of power, knowledge and resources"[20].

Counter-narratives of smart urbanism can be produced through "speculative prototyping"[21], or be initiated through grassroot movements, as in the case of Barcelona, where citizens, academics and new political leaders have put forward the idea of "technology sovereignty".[22] A number of experiences in Barcelona suggest that the success of smart initiatives depend not on sensors and data but on the development and implementation of community building skills and on the production of data that is truly meaningful and valuable for the citizens.[23] These variegated forms of engagement with smart cities raise important questions about the forms of public engagement that were made possible during the robot competition in Milton Keynes. First, however, we look at how and why robot competitions started to figure in the European Commission's plans.

4. Institutionalizing European Robot Competitions

Robot competitions have existed for decades.[24] However, the vision of smart cities has extended their visibility beyond the specialist circles that thus far have been their main audience and participants. Typically organized around a clear thematic focus, – playing football, destroying an opponent robot, picking up items from a shelf – competitions allow developers and programmers to think about the specific tasks that performing an action involves and about the environmental elements that the robot must be able to identify: mobile and immobile objects, walls, navigable surfaces, obstacles, voice commands, objects to grasp, and so on. In one of the first and most popular competitions, *RoboCup*, for example, participants program robots to play football: to run behind a ball, make or block passes, reach the opponents' goal and shoot the ball out of the reach of a goalkeeper robot. All these actions involve

19   Dorien Zandbergen/Justus Uitermark, In Search of the Smart Citizen: Republican and Cybernetic Citizenship in the Smart City, in: *Urban Studies* 57 (8/2020), 1733–48.

20   Helen Manchester/Gillian Cope, Learning to Be a Smart Citizen, in: *Oxford Review of Education* 45 (2/2019), 224–41, see 224.

21   Martín Tironi, Prototyping Public Friction: Exploring the Political Effects of Design Testing in Urban Space, in: *The British Journal of Sociology* 71 (3/2019), 1–17; Martín Tironi, Speculative Prototyping, Frictions and Counter-Participation: A Civic Intervention with Homeless Individuals, in: *Design Studies* 59 (2018), 117–38.

22   Evelien de Hoop et al., Smart Urbanism in Barcelona: A Knowledge-Politics Perspective, in: Jens Stissing Jensen/Matthew Cashmore/Philipp Späth, *The Politics of Urban Sustainability Transitions*, London 2018; Smith, Smart Cities Need Thick Data.

23   Smith, Smart Cities Need Thick Data.

24   RoboCup Federation, Official Website, URL: https://www.robocup.org/ [last accessed: August 15, 2023].

complex sequences of navigation, motion, and visualization that have to be patiently and carefully programmed and integrated.

Besides putting the skills of the programmers and the reliability of robotic platforms to test, robot competitions aim to foster education, team development, and public engagement with science and technology. They also aim to facilitate open innovation. In recent years, a wide range of organizations have engaged in the practice of assembling robot competitions. Prominent examples include the DARPA (Defense Advanced Research Projects Agency, US) Robotics Challenge (2012–2015); the ARGOS (Autonomous Robot for Gas & Oil Sites) Challenge, organized by the French TOTAL (2013–2017), and the Mohammed Bin Zayed International Robotics Challenge (MBZIRC), which Khalifa University of Science and Technology organizes since 2017 in Abu Dhabi. The European Commission itself started to fund robot competitions in 2013.[25]

Although all competitions have to some extent educational, entertainment, scientific, and a problem-solving value, it is possible to find significant variations when the organizers and sponsors are transnational oil and gas corporations, military organizations, academic communities, or supranational entities like the European Union. Different organizers imagine in different ways what robotic technologies are for and what robotic futures should be brought into being, and how. However, significant overlaps exist between the organizers, advisors, and participants of different competitions. Thus, there is a complicated nest of institutional structures that give shape to these competitions.

According to one of the main actors of the ERL, officers from the EC became interested in robot competitions after they were invited to the 2009 edition of *RoboCup* in Graz, Austria. The potential they saw in competitions for education, dissemination, and public engagement encouraged them to include calls dedicated to fund robot competitions in the European innovation strategy. From 2013 to 2015, the Commission funded *RockIn*, which organized tournaments in the scenarios of industrial robots (RockIn@Work) and the home environment (RockIn@Home); and *Eurathlon*, which focused on emergency and rescue robots. These competitions imported a number of features from *RoboCup*, for example, dividing the competition in different areas of application and applying test benchmarks.

The European competitions continued receiving support from the EC from 2016 to 2018 through *EuRoC* and *RockEU2*. These initiatives contributed to the foundation of the ERL and to bring all competitions under a single institutional entity. This merger enabled the establishment of common ground between different robotics

25   CORDIS 2017a, Robot Competitions Kick Innovation in Cognitive Systems and Robotics, RoCKIn Project | FP7| CORDIS | European Commission, April 22, 2017, URL: https://cordis.e uropa.eu/project/id/601012 [last accessed: August 15, 2023].

communities, but it also revealed conflicting understandings of the aims and rationale of robot competitions. To give an example, the organizers of competitions in the home and industrial environments intended to run competitions in standardized environments. However, for the organizers of the emergency and rescue competitions, a standardized environment had little value because emergency and rescue missions occur in random and chaotic environments, where conditions such as light, wind, and humidity are out of human control.

SciRoc was the successful grant application to a call from the EC that aimed to increase public understanding of robotics, assess public perception of robotics, strengthen the collaboration between diverse robotics communities, and increase public and private investment in robotics development through competitions[26]. *SciRoc* received funding to organize two biennial competitions, in 2019 and 2021, and other related public engagement and dissemination activities. The consortium was formed by 8 universities from the UK, Spain, Italy, Portugal and Germany, the NATO Science and Technology Organization, and euRobotics, the organization that coordinates the development of the Europe robotics community.[27] Besides the Milton Keynes competition, in January 2021 the project website announced that the second edition of *SciRoc* would take place in Bologna.

It was not a coincidence that the first edition of *SciRoc* took place in Milton Keynes. Not only was the Open University, a consortium member, based at Milton Keynes, but also, during previous years, the city council had invested efforts and resources in re-inventing itself to become more attractive to foreign investors and to portray itself as a smart city.[28] The city council established in 2015 the "MK Futures 2050 Commission", which was in charge of developing a program for the future of the city. The program included measures such as the creation of a new technological university with a strong focus on digitalization, a "new vision for the city centre, and a smart city program related to intelligent and autonomous mobility"[29]. The city also had recently welcomed the implementation of grocery delivery robots from the Estonian company *Starship Technologies*, which the citizens found useful even

26   CORDIS 2017b, "Robotics Competition, Coordination and Support." October 31, 2017. https:// cordis.europa.eu/programme/id/H2020_ICT-28-2017 [last accessed: August 15, 2023].

27   Kjetil Rommetveit/Niels van Dijk/Kristrún Gunnarsdóttir, Make Way for the Robots! Human- and Machine-Centricity in Constituting a European Public–Private Partnership, in: *Minerva* 58 (1/2020), 47–69; CORDIS 2021, "European Robotics League plus Smart Cities Robot Competitions." February 25, 2021. https://cordis.europa.eu/project/id/780086 [last accessed: August 15, 2023].

28   Alan-Miguel Valdez/Matthew Cook/Stephen Potter, Roadmaps to Utopia: Tales of the Smart City, in: *Urban Studies* 55 (15/2018), 3385–3403.

29   Jeremy Coward, Why Milton Keynes Is One of the Smart Cities in the World, in: *IoT World Today*, 16.04.2018, URL: https://www.iotworldtoday.com/2018/04/16/why-milton-keynes-one-smart-cities-world/ [last accessed: August 15, 2023].

before the Covid-19 pandemic.[30] The vision of a smart and digitalized Milton Keynes facilitated the collaboration and coordination between the *SciRoc* organizers, the city council, and other local organizations.

5. Three Modes of Human-Robot-Public Engagement

In the following section, we provide a description of how the *SciRoc* competition unfolded. Emerging from a grounded and iterative process of comparing empirical material with ideas from the literature, we identified three modes of human-robot-public engagement: embracing engagement, bypassing engagement and prefiguring engagement. We organize our discussion of the empirical material by re-constructing a case narrative that makes sense of these three types of engagement before offering discussing the implications for the shaping of human-robot-public engagement in smart cities.

5.1   Embracing Engagement

SciRoc was driven by many motivations. Some of the organizers saw the event as an opportunity to increase public and democratic participation in the development of digital and robotic technologies. Months before the event, one of the organizers we talked to claimed that the main goal of the event was to get people informed about where the technology is going, to let them ask questions, and to make it possible to have a discussion. This, he said, so that decisions about what futures of robotics and artificial intelligence are and are not desirable are made collectively rather than by a few Big Tech companies like Google and Amazon. The purpose of these public competitions, according to our informant, was to explore "*how robots 'respond' to society and how society responds to the robots*". Furthermore, he argued that an additional goal of hosting robot competitions as public events in realistic environments was to search for ways to make the ERL economically sustainable, so that it could stop relying on funding from the European Commission.

In order to attract a wide range of participants and sponsors, the competition arena offered five different scenarios or "episodes" that illustrated different tasks that robots could carry out in a smart city. These included: "Deliver coffee shop orders, Take the elevator, Shopping pick and pack, Open the door, and Fast delivery

---

30   Alex Hern, Robots Deliver Food in Milton Keynes under Coronavirus Lockdown, in: *The Guardian*, April 12.04.2020, URL: http://www.theguardian.com/uk-news/2020/apr/12/robo ts-deliver-food-milton-keynes-coronavirus-lockdown-starship-technologies [last accessed: April 4, 2024].

of emergency pills"[31]. *Costa* was the sponsor of the coffee shop scenario. This resembled a usual coffee shop from the British chain and had capacity for five tables and a cashier desk (see Fig. 1). The space between the tables was wide enough so that the robot could navigate easily. To the left side located the scenario of the "Fast delivery of emergency pills" episode. A net cage was built in which drones could fly across obstacles to come close to a real-size human dummy in need of a first aid kit (see Fig. 4). On the back of the cage for drones located the "Shopping and pick and pack" scenario. This was a more simple scenario that consisted of a desk with shopping baskets on top and a few shelves a few meters back. The technology and grocery retailer and delivery company *Ocado* was the sponsor of this episode (see Fig. 2). Behind the *Costa* situated the "Take the elevator" scenario, or rather a foyer with two silver colored sliding doors (that opened manually) that resembled two elevators, with rooms the size of an ordinary elevator behind them (see Fig. 3). Behind the "Shopping and pick and pack" scenario located the "Open the door" scenario, which resembled the waiting room for an office or the corridor outside of an apartment.

The sponsors added a different layer of rules and interests to the nest of institutional values that brought the competition into being. While *Costa* benefited mainly from the publicity, *Ocado* also contributed to the design of the "Shopping pick and pack" episode. To do so, it provided ideas to develop the ERL benchmarks according to what the company considered relevant for food packing. For example, including a more strict timing, and identifying and grasping items of different shapes and sizes. Moreover, by sharing their usual activities and their logos, these sponsors contributed to bring the unfamiliar world of robot applications into a more familiar context for the audience.

The competing teams were mainly students from four British universities and single universities from Germany, Portugal, Spain, and France. A team from the Spanish branch of the Japanese telecommunications company NTT also participated. The different episodes gave teams different opportunities to try out their existing programming skills or develop new ones. Some teams focused on drone navigation, others in grasping and visualization, and others in robot navigation, voice recognition, and computer vision. The stage and the possibility to program a robot to perform in a highly populated environment made the competition appealing for teams with different skills and backgrounds. Programming a robot to carry out apparently simple tasks involves many hours of work in front of the computer. Making a robot move, react to voice commands, respond back, and grasp objects requires at least one person to be in charge of each of these different tasks. Competitions offer a caring environment in which university students and

---

31 Damian Dadswell, First *SciRoc* Challenge 2019 (3.08.2018b), URL: http://instituteofcoding.op en.ac.uk/ [last accessed: August 15, 2023].

more experienced programmers can develop their skills in a relatively quiet and undisturbed place.

Not many universities have the capacity to support a team for robot competitions. University teams have limited resources, therefore they have to select carefully the competitions they would like to attend every year. Having a robot and transporting it is costly. In order to make it possible for a larger number of teams to participate in *SciRoc*, Barcelona based *PAL Robotics* leased their popular model *Tiago* (Take-it-and-go) at a low cost for teams that did not own a robot or were not able to bring their own. *PAL Robotics* also brought a number of technicians to provide support to the teams whenever their *Tiagos* were not responding as expected.

Other sponsors and organizations that increased the visibility of the event had exhibition stands on the back of the competition arena. Amongst them figured Cranfield University, the University of West of England, the Open University, PAL Robotics, Vodafone, Westcott 5G Step-Out Centre, the city council, a local engineering network, a national innovation networking initiative called Catapult, and a few others. Besides encouraging social acceptance, the event was also a way of making higher education more attractive for individuals who may not have considered going into it. Moreover, the non-academic institutions, by contrast, provided evidence that technologies not only would make a great impact on daily life but also generate jobs and build a new and more prosperous Milton Keynes.

One group that had the greatest chances to engage with the robots and the event were about 50 pre-selected volunteers from a local network of engineering professionals, technicians, and students. These volunteers interacted with the robots during the coffee shop and the elevator episodes. In the coffee shop, they made an order and took it from the tray when the robot brought it from the cashier desk. In the elevator, volunteers reacted to the robots' request to press a certain number. The volunteers were asked to fill questionnaires to assess their interaction with the robots. In this way, volunteers represented the most direct and explicit way in which society was brought to the robots, and the robots to society. The ERL opted for pre-selected volunteers to have them better informed about the situation and avoid any accidents, but this was also a way to have more control over how the society could engage with the event.

Besides the main event and the exhibitions in the shopping mall, a series of events were held in the offices of the city council and the public library, not far from there, during the week of the competition. The inaugural event included an expert panel discussion followed by a reception with wine and canapés. The audience was invited to ask questions which were addressed by the experts. Answering one question, the Open University Vice-Chancellor argued that these events and the advertisement of higher education it involved had explicit democratic ends. He stated that *"you can only democratise technology if you have an educated population"*. He further argued that *"the question is whether you want to program or to be programmed"*, thus calling for a direct and active involvement of the population in the digital transformation of society.

The idea of engaging with the public was more actively embraced in informal conversations around the SciRoc arena in the shopping mall. In particular, one of the organizers and a representative from the city council spent a substantial amount of time walking around and talking to the audience, making them see what they saw in the event, expressing what they thought the deeper meanings of these competitions were, and listening to the thoughts from people from the audience. But these encounters were rather unusual. Only in the last days of the event volunteers and a few competitors got involved in this practice. The organizers who knew us beforehand asked us to go around and talk to people about what was going on there. As ethnographic researchers, we didn't need to be asked twice.

There were members of the audience who had seen the event and had later come back with their children. One of them said that she was interested in exposing her child to what it takes to program a robot, and possibly so that he could see how robots are built internally. Yet, she also questioned *"why are we investing in robots to do these jobs and produce more, if the population is growing anyway?"* Leila, the student we described in the introduction to this chapter, said it took her a while and multiple visits to understand that this was a competition. Moreover, she wondered how deaf people or non-English speakers could make sense of the event. She also wondered how the robots could address their needs, pointing a finger to the fact that the competition, besides all the efforts to promote public engagement with a roboticized future, was nevertheless oriented towards a limited audience.

5.2   Bypassing Engagement

Although *SciRoc* was oriented to engage publics with science and technology, different features of the setting and situation played against this goal. For the participants, this was a much more stressful environment than other local tournaments we attended in Oldenburg and Bristol earlier that year. In those competitions, only one team participated in each to have their scores compared later. While in those local tournaments participants had between three and five days to program and practice in standardized environments inside the lab without interference with other teams, during *SciRoc* the teams had little time to rehearse and to repeat their performances.

Moreover, the internet connection was weaker than back in the labs, and the robots were disturbed by the changes of light and the numerous unrecognized faces of competitors, volunteers, and the audience at the other side of the fence. Additionally, the noise made it difficult for the robot to recognize voice commands. In the case of the drones delivering the first aid kit, the wi-fi signal was so weak – due to the hundreds of mobile phones connected to the network – that they could barely perform. In the *Costa* café episode, team members had to request the volunteers to look directly at the robot's vision cameras and speak loud and clear to make an order. In that same episode, while we were looking from the inside of the stage and standing just behind the 1.20m high hoardings, we were asked to move some steps back because the robot could not distinguish the persons sitting at the table, who it should attend, from those who it should ignore.

The technical challenges of the competition and the aim to have a good performance meant competing participants were keen to avoid being distracted by the public. During a general meeting, a participant asked in a reluctant tone if they were expected to respond to inquiries from the public. One of the organizers responded that they could try *"to evade the questions, send them somewhere else, or just ignore them"*. On another occasion, a member of the audience was asking several questions to one of the participants about why to use grips rather than suction cups for the "Shopping picking and packing" episode. When the team member tried to answer, his teammates called him, annoyed and desperate, to get back to his position. The tension between holding a serious competition and facilitating the engagement between publics, robots, and experts was also reflected in discussions regarding the space that should be left between the queuing belts and the borders of the arena. Some of the organizers wanted to have these belts less than a meter closer so that the audience could get a better look. However, some of the competitors and other organizers wanted to keep them at least one meter further to avoid distractions.

The technical challenges that the competing teams faced meant that the actual performances went very slow. The time that it took for one single team to bring their robot to the starting point of an episode, for example, in the *Costa* café scenario, and making it supervise tables, take an order and bring items to the customers/volunteers, took about twenty minutes. The time between one team leaving and another one coming in was five to ten minutes. In this way, if the audience wanted to see the robots perform, they needed to be extraordinarily patient. The volunteers, sitting at the café tables inside the competition arena, also had to wait patiently.

The setup of the stage also contributed to make it hard for the public to engage with the event. In particular, there were no chairs, benches, high tables, or handle rails where the audience could relax while waiting for the robots to appear and perform. People had to wait standing in the middle of corridors for undetermined time. Many of the attendants were carrying their children, grocery shopping bags and other items, therefore they were not keen to wait for too long.

Making sense of the unfamiliar situation that the competition brought forward required a large amount of curiosity, time, and patience from the audience. One attendant who worked in the IT sector and who would count as an informed member of the audience, argued that it took him a few rounds to the stage and some 10 to 15 minutes to figure out what was going on there. However, he could not tell where the teams were from or how many they were, and neither what the scoring consisted of. Many of the people passing by would not slow down to try to make a careful reading of the event. Those who would stop for a moment would continue walking as soon as failure occurred. A group of women that looked for a little while at a team competing on the "Fast delivery of emergency pills" episode, turned around saying *"booooys with their tooooys"*, before walking away, as soon as the drone hit an obstacle and dove to the ground.

There were members of the audience who opted for not listening even when somebody – most likely one of us – offered some orientation or intended to start a conversation. A group of elderly people argued *"well, at least it's more polite [than human staff]"* when they heard the robot taking an order. Another member from the audience argued that automation, for example in the electronic cashiers at the supermarket, are *"already taking people's jobs"* even though they *"fail all the time'*.

Finally, the complexity of the different episodes, scenarios, tasks, and teams that the competition involved meant that the audience did not have access to many of its features. To give an example, in the *Costa* café episode, the robot had to be programmed to autonomously (a) go to a table, (b) ask the volunteers if they were already being served, and if not, (c) to take their order, direct itself towards the cashier desk and request the items ordered. Then, (d) the robot had to identify if the items placed on a tray were the right ones, and if so, (e) bring them to the table that made the order. Since the cashier desk was far from the audience, they could not see or hear what (d) was about. Furthermore, on a few occasions, the buzzing sound produced by the drones made it difficult to listen to what was happening in the other scenarios, creating an unpleasant situation not only for the audience but most likely also for the competing participants.

5.3   Pre-figuring Engagement

The tension between *SciRoc*'s objectives of, on the one hand, holding a complex competition that was appealing for the robotics community, and on the other hand engaging with publics, deserves further attention. *SciRoc* built on the more than five years of experience of the ERL organizing competitions. For some members of this community, *SciRoc* represented a way of making competitions more challenging by putting them in a more realistic environment and increasing the complexity of the human-robot interaction challenges. However, the more was added to the already sophisticated scoring system and rationale, the harder it was for the lay audience to get a sense of what was going on and to engage more meaningfully with the event.

SciRoc became a complex, sophisticated, and multilayered event because it was the continuation of a series of commitments that were deeply ingrained in the institutional values of the ERL. To begin with, the five different scenarios that the competition offered built on and derived from the existing focus areas funded through *RockIn* and *Eurathlon* in previous years: industry, home, and emergency and rescue robots. The breadth of the notion of smart cities made it possible for the ERL to bring the three existing competition scenarios under a single and more coherent narrative. But most importantly, the ERL values were to a large extent shaped by the expectations and grant conditions of the European Commission.

To be more appealing for the European Commission, the ERL had to become more compatible with the Commission's interests, including the creation of a "European identity". According to ERL representatives, the difference between the ERL and other competitions is that the former are more explicitly focused on human-robot interaction and on assessing robot performance through scientific benchmarking.[32] In addition, these competitions addressed "current European challenges" that resonated with other focal areas that the Commission had funded such as the digitalization of industries and the growing aging population. In this way, the focus on European challenges, scientific benchmarking, and human-robot interaction, provided the ERL competitions an "European flavor", to use the words of one of our informants, that made them distinctive.

Besides educational, dissemination, and competitive purposes, the ERL competitions aimed at experimenting with, and advancing, the standardization of "benchmarks" through which robots and programmers' skills could be assessed "objectively".[33] For that purpose, the ERL and partner institutions organized local and major tournaments in different locations across Europe. These were facilitated by the establishment of a number of standardized environments, or "certified test beds", where the competitions could take place. Starting with a test bed for industrial robots at the Bonn-Rhein-Sieg University of Applied Science (Germany) and a test bed for robots for the home environment in the University of Lisbon (Portugal) during the project *RockIn*, certified test beds were installed in academic robotics laboratories in Edinburgh (UK), Bristol (UK), Leon (Spain), Peccioli (Italy), Oldenburg (Germany), and in the headquarters of *PAL Robotics* in Barcelona (Spain).

In these test beds, competitions were held offering the same rules, tasks, and scores. Having multiple sites for competitions was envisioned as a way of encouraging teams to participate in more than one event and thus enabled a more reliable and statistically significant assessment of their performance. What is more, since competitions encouraged the mobility of participants between different European countries hosting competitions, the ERL promoted the formation of a European identity. To one of our informants, this consisted of making young students

---

32   Pedro U. Lima et al., RoCKIn Innovation Through Robot Competitions [Competitions], in: *IEEE Robotics Automation Magazine* 21 (2/2014), 8–12; Francesco Amigoni et al., Competitions for Benchmarking: Task and Functionality Scoring Complete Performance Assessment, in: *IEEE Robotics Automation Magazine* 22 (3/2015), 53–61; Sven Schneider et al., Design and Development of a Benchmarking Testbed for the Factory of the Future, in: *2015 IEEE 20th Conference on Emerging Technologies Factory Automation (ETFA)*, Luxembourg 2015, 1–7.

33   Lima et al., RoCKIn Innovation; Amigoni et al., Competitions for Benchmarking; Schneider e t al., Design and Development of a Benchmarking Testbed.

aware of their proximity to other European countries and the possibility of forming part of the same community. This, he suggested, could help to mitigate the current growth of right-wing sentiments across Europe.

With exception of the certified test bed in Bonn for the industrial scenario, the certified test beds resembled an apartment for a lone elderly person, possibly in a care home, consisting of half walls and *Ikea*-like furniture, a kitchen, a living room, a bedroom, and a few doors. To situate the competitions into a more realistic context and to make an emphasis on the sociability of robots, the rulebook of the *ERL Consumer Service Robots* provided the following *Consumer User Story*:

> Granny Annie is an elderly person, who lives in an ordinary apartment. Granny Annie is suffering from typical problems of ageing people: She has some mobility constraints. She tires fast. She needs to have some physical exercise, though. She needs to take her medicine regularly. She must drink enough. She must obey her diet. She needs to observe her blood pressure and blood sugar regularly. She needs to take care of her pets. She wants to have a vivid social life and welcome friends in her apartment occasionally, but regularly [...] For all these activities, ERL Consumer is looking into ways to support Granny Annie in mastering her life.[34]

In this way, the certified test beds contributed to reproduce and lock-in a dominant vision that robotic technologies play a key role in "ambience assisted" and "independent" living of elderly people. They also contributed to lock-in a particular vision of human-robot interaction.

In earlier competitions participants had to program robots to carry out a number of given tasks (or "episodes") oriented to cater for *Granny Annie*. These included: receiving guests and distinguishing familiar and unfamiliar faces, asking who the guests are and responding in different ways by providing different instructions; navigating the apartment autonomously, detecting new changes in the environment (e.g. by recognizing newly added items or misplaced furniture), ideally being able to manipulate objects (e.g. a cup, a TV remote control) and bringing them back to the right place. The teams received points depending on how optimally these actions were carried out, and penalized if the robot accidentally hit elements of the environment, or people. In these competitions, nobody embodies the persona of *Granny Annie*, and nobody besides the roboticists contribute to the evaluation of the teams, or of the tasks themselves.

During *SciRoc*, most of these challenges were exported from this scenario to the smart city context. Many of them became part of the episode at *Costa* café, the opening the door challenge, and the shopping pick and pack challenge. In this way, the

---

34   Meysam Basiri/Pedro U.Lima, European Robotic League for Consumer Service Robots, URL: https://www.eu-robotics.net/robotics_league/upload/documents-2018/ERL_Consume r_10092018.pdf [last accessed: August 15, 2023].

highly controlled environment of the lab was brought into the shopping mall in Milton Keynes; and with it, a particular way of envisioning human-robot interaction. This, however, left not much room for designing a competition that gave more active and engaging roles to the audience – or to the pre-selected volunteers.

6. Discussion: Shaping Social Orders in the Smart City

So, what does this narrative tell us about how the robot competition engaged with publics and shaped social orders in the smart city? Our narrative foregrounds three different modes in which the competition, its organizers, and sometimes its publics dealt with the notion of engagement. First, *embracing engagement*, an open and attentive form of engagement that was sensitive to the needs, interests, and concerns of various participants, sponsors, and members of the audience. Second, *bypassing engagement*, a more constrained *and* constraining form of engagement that limited the possibilities of mutual understanding between competition participants and the various publics. Third, *prefiguring engagement*, a variety of previous commitments and expectations that brought the event into being and gave it shape, but that rigidly framed the ways in which publics and participants could engage with each other.

These three modes of engagement in turn revealed and were shaped by different logics of social ordering. Here we trace three logics, namely conviviality, control, and care, which were differently articulated in each mode of engagement, and in different amounts. We expand on these briefly to better illustrate the interplay between these modes of engagement and social ordering.

6.1   Conviviality

Immediately upon entering the mall, the competition sought to project to visitors and participants alike a sense of conviviality. In the competition, conviviality between humans and robots was suggested in the poster illustrations and in the competition episodes or scenarios. Episodes were scripted to reward convivial and harmonious interactions – at least within the borders of the competition arena. But conviviality is more than being together in harmony. Conviviality describes an understanding of people and things in society that is – in principle – relational, mutualistic, and egalitarian.[35]

The establishment of convivial relations was best exemplified in the work of embracing engagement that a lead organizer and a city council representative did walking the competition arena barrier. This boundary work was critical in members of the robotics community gaining respect for their host public. They actively solicited

---

35   Ivan Illich, *Tools for Conviviality*, London 1973.

engagement, generously making time for shoppers as well as offering insight and sharing information. This was work of communication and translation in both directions as they conveyed sentiment in almost real time back over the barriers to competitors.

Nevertheless, mutualistic engagement that fostered deep understanding was rare. Current social theory examines conviviality through autonomy and self-realization, as a logic that resists technocratic control and coloniality and opens up possibilities for different kinds of future social orders. A convivial society is one in which social and material interdependencies are political and mediated by tools, institutions and practices across that society.[36] But this kind of expansive conviviality was not up for grabs in Milton Keynes. Ultimately there was little opportunity for collectively imagining a future smart city that wasn't already scripted in the competition episodes.

A defining characteristic of conviviality for Illich was its imperative against technologies and tools that seek to control and dominate – a problem of centralizing institutions and structures associated with modern industrialization.[37] And yet, some of the foundational sociological work critiquing the smart city emphasizes how surveillance technologies and similar infrastructural and institutional arrangements constitute a technocratic logic of control.[38] Social control can also be achieved through exclusion as well as participation. For instance in experimental settings that foster creativity, participation and innovation such as makerspaces, gender, class and race have been shown to structure who gets to participate and who doesn't.[39]

6.2   Control

We observe a logic of control – in the first instance – on how borders and boundaries were established between the competition arena and the ordinary corridors of the shopping mall around it. The barriers were there to let the competition occur without interruptions from the audience that the organizers considered unnecessary. In this way, multiple control mechanisms made it possible for participants to bypass engagement. The barriers, however, contributed to make the arena identifiable and suggested interpretative framings for the audience, thus they were about control but also about embracing engagement.

---

36   Saurabh Arora et al., Control, Care, and Conviviality in the Politics of Technology for Sustainability, in: *Sustainability: Science, Practice, and Policy* 16 (1/2020), 247–62.

37   Illich, Tools for Conviviality.

38   Orit Halpern et al., Test-Bed Urbanism, in: *Public Culture* 25 (2/2013), 272–306; Jathan Sadowski/Frank Pasquale, The Spectrum of Control: A Social Theory of the Smart City, in: *First Monday* 20 (7/2015).

39   O'Donovan/Smith, Technology and Human Capabilities.

Control was also the underlying logic behind scoring mechanisms and scientific benchmarking that gave meaning to the ERL competitions, thus control was fundamental for prefiguring engagement.

The logic of control was also present in the role of the volunteers. Their participation enabled the organizers to bring in non-specialist societal actors into the competition – a form of embracing engagement –; but in contrast the volunteer roles made no room for less prescribed forms of interaction between volunteers, robots, and competing roboticists – a form of bypassing engagement. In the passive and cooperative roles that the organizers planned for the volunteers there was little room to imagine more complex and ambivalent human beings who may have gone into higher education to study computer science, but who then opted for a traditionally gendered occupational training, such as Leila.

The figure of the volunteers, however, played a key role in pre-figuring engagement because their participation increased the controlled complexity of the human-robot interaction component of the competition.

6.3   Care

By comparison with control, logics of care direct attention to neglected things and devalued doings.[40] For instance, the hidden labors of care workers[41], or marginalized groups excluded from social services. *Granny Annie* figured in previous competitions as the character who made visible the usually unnoticed challenges of elderly people living alone. In the robotics laboratories we visited for earlier competitions in Oldenburg and Bristol we learned about procedures from the researchers for engaging directly with members of the public, involving them in forms of co-creation and co-design, methods of participative innovation.[42] However, while the standardized benchmarks and tasks such as navigation and grasping objects were imported to *SciRoc*, the image and values that *Grannie Annie* evoked were not too visible in the smart city vision that the competition enacted.

A logic of care was also present in the engagement activities within and around the arena. Senior researchers took seriously the responsibility of fostering interdisciplinary capabilities in the competitors – often committing to team building mentoring and management over many years. Furthermore, one side of the arena's boundary was filled with stalls representing partner universities. For some of the senior researchers, (embracing) engagement meant not only initiating recruitment

40   María Puig de la Bellacasa, *Matters of Care: Speculative Ethics in More than Human Worlds. Matters of Care: Speculative Ethics in More than Human Worlds*, Minneapolis 2017.

41   Peter A. Lutz, Surfacing Moves: Spatial-Timings of Senior Home Care, in: *Social Analysis* 57 (1/2013), 80–94.

42   Michalec/Sobhani/O'Donovan, What Is Robotics Made of?

conversations with passers-by, but also communicating research and outreach op-portunities to participate with universities in different ways.

7. Conclusion

To summarize our discussion: we have mobilized ideas about conviviality, care and control as specific logics of social ordering extant in the literature. This of course is not to say that they are the only orderings of people, things and knowledge in the story. Indeed, the point that we have tried to make is that each are visible in differ-ent parts of the situation. Some are brought into the situational maps via processes of path dependence and contingencies (the operating systems that power the Tiago robots for instance), while other examples come about through the choices people make in the moment. This tells us that there is not one singular robotic smart city framing being enacted. Rather, many smart cities are possible.

The situation of the robot competition in Milton Keynes that we opened up in this chapter made salient a number of themes that are commonplace in organiza-tion studies and institutional theory literature. The competition itself drew strongly on the path dependence[43] established by the ERL over previous years, which in turn derived from the use of competitions in different environments for the sake of in-novation. The organizers could not abandon the intrinsic elements that made the competition scientifically complex precisely because those had been the reason they received funding in the first place. Put differently, neither robots nor interaction ar-rived value free but already carried the baggage, values, trajectories, pre-existing commitments of the organizers. Yet, feel encouraged to conclude that making a sim-pler competition could have contributed to forms of embracing rather than bypass-ing engagement with publics. A simpler competition could have also emphasized the logics of care and conviviality in different ways rather than the logic of control.

Our study also highlights the risks of reproducing controlling mechanisms in-trinsic in mainstream notions of the smart city rather than searching for more car-ing and convivial alternatives. These, in turn, run the risk of restricting the chances to imagining and institutionalizing other futures of human-robot interaction.

---

43   Jorg Sydow/Georg Scheyogg/Jochen Koch, Organizational Path Dependence: Opening the Black Box, in: *Academy of Management Review* 34 (4/2009), 1–21.

Bibliography

Amigoni, Francesco/Bastianelli, Emanuele/Berghofer, Jakob/Bonarini, Andrea/ Fontana, Giulio/Hochgeschwender, Nico/Locchi, Luca et al., Competitions for Benchmarking: Task and Functionality Scoring Complete Performance Assessment, in: *IEEE Robotics Automation Magazine* 22 (3/2015), 53–61.

Arora, Saurabh/Van Dyck, Barbara/Sharma, Divya/Stirling, Andy, Control, Care, and Conviviality in the Politics of Technology for Sustainability, in: *Sustainability: Science, Practice, and Policy* 16 (1/2020), 247–62.

Basiri, Meysam/Lima, Pedro U., European Robotic League for Consumer Service Robots, URL: https://www.eu-robotics.net/robotics_league/upload/documents -2018/ERL_Consumer_10092018.pdf [last accessed: July 30, 2021].

Baykurt, Burcu/Raetzsch, Christoph, What Smartness Does in the Smart City: From Visions to Policy, in: *Convergence* 26 (4/2020), 775–89.

Bellacasa, María Puig de la, *Matters of Care: Speculative Ethics in More than Human Worlds. Matters of Care: Speculative Ethics in More than Human Worlds*, Minneapolis 2017.

Bulkeley, Harriet/Castán Broto, Vanesa, Government by Experiment? Global Cities and the Governing of Climate Change: Government by Experiment?, in: *Transactions of the Institute of British Geographers* 38 (3/2013), 361–75.

Clarke, Adele E., *Situational Analysis: Grounded Theory After the Postmodern Turn* (1st edition), Thousand Oaks 2005.

CORDIS. 2017a, Robot Competitions Kick Innovation in Cognitive Systems and Robotics, RoCKIn Project | FP7 | CORDIS | European Commission, April 22, 2017, URL: https://cordis.europa.eu/project/id/601012 [last accessed: August 15, 2023].

CORDIS. 2017b. "Robotics Competition, Coordination and Support." October 31, 2017. https://cordis.europa.eu/programme/id/H2020_ICT-28-2017 [last accessed: August 15, 2023].

CORDIS 2021. "European Robotics League plus Smart Cities Robot Competitions." February 25, 2021. https://cordis.europa.eu/project/id/780086 [last accessed: August 15, 2023].

Coward, Jeremy, Why Milton Keynes Is One of the Smart Cities in the World, in: IoT World Today, 16.04.2018, URL: https://www.iotworldtoday.com/2018/04/16/wh y-milton-keynes-one-smart-cities-world/ [last accessed: August 15, 2023].

Dadswell, Damian, ERL Smart Cities (2018a), URL: http://instituteofcoding.open.a c.uk/ [last accessed: August 15, 2023].

Dadswell, Damian, First *SciRoc* Challenge 2019 (3.08.2018b), URL: http://instituteof coding.open.ac.uk/ [last accessed: August 15, 2023].

Davies, Sarah R., An Empirical and Conceptual Note on Science Communication's Role in Society, in: *Science Communication* 43 (1/2021), 116–33.

Engels, Franziska/Wentland, Alexander/Pfotenhauer, Sebastian M., Testing Future Societies? Developing a Framework for Test Beds and Living Labs as Instruments of Innovation Governance, in: *Research Policy* 48 (9/2019), 1–25.

Halpern, Orit/Lecavalier, Jesse/Calvillo, Nerea/Pietsch, Wolfgang, Test- Bed Urbanism, in: *Public Culture* 25 (2/2013), 272–306.

Hern, Alex, Robots Deliver Food in Milton Keynes under Coronavirus Lockdown, in: The Guardian, April 12.04.2020, URL: http://www.theguardian.com/uk-news/2 020/apr/12/robots-deliver-food-milton-keynes-coronavirus-lockdown-starshi p-technologies [last accessed: April 4, 2024]

Hoop, Evelien de/Macrorie, Rachel/Smith, Adrian/Marvin, Simon, Smart Urbanism in Barcelona: A Knowledge-Politics Perspective, in: Jens Stissing Jensen/ Matthew Cashmore/Philipp Späth, *The Politics of Urban Sustainability Transitions*, London 2018.

Illich, Ivan, *Tools for Conviviality*, London 1973.

Jasanoff, Sheila, *States of Knowledge: The Co-Production of Science and the Social Order*, London, New York 2004.

Laurent, Brice/Doganova, Liliana/Gasull, Clément/Muniesa, Fabian, The Test Bed Island: Tech Business Experimentalism and Exception in Singapore, in: *Science as Culture* 30 (3/2021), 367–90.

Lima, Pedro U./Nardi, Daniele/Kraetzschmar, Gerhard/Berghofer, Jakob/ Matteucci, Matteo/ Buchanan, Graham, RoCKIn Innovation Through Robot Competitions [Competitions], in: *IEEE Robotics Automation Magazine* 21 (2/2014), 8–12.

Lutz, Peter A., Surfacing Moves: Spatial-Timings of Senior Home Care, in: *Social Analysis* 57 (1/2013), 80–94.

Macrorie, Rachel/Marvin, Simon/While, Aidan, Robotics and Automation in the City: A Research Agenda, in: *Urban Geography* 42 (2/2019), 1–21.

Manchester, Helen/Cope, Gillian, Learning to Be a Smart Citizen, in: *Oxford Review of Education* 45 (2/2019), 224–41.

Marvin, Simon/While, Aidan/Kovacic, Mateja/Lockhart, Andy/Macrorie, Rachel, *Urban Robotics and Automation: Critical Challenges, International Experiments and Transferable Lessons for the UK*, EPSRC UK Robotics and Autonomous Systems (RAS) Network.

Michalec, Ola/Sobhani, Mehdi/O'Donovan, Cian, What Is Robotics Made of? The Politics of Interdisciplinary Robotics Research, in: *Humanities & Social Sciences Communications* 8 (2021), article 65.

Mosco, Vincent, *Smart City in a Digital World*, Bingley 2019.

Muggah, Robert/Walton, Greg, Smart' Cities Are Surveilled Cities, URL: https://for eignpolicy.com/2021/04/17/smart-cities-surveillance-privacy-digital-threats-i nternet-of-things-5g/ [last accessed: August 15, 2023].

O'Donovan, Cian, Accountability and Neglect in UK Social Care Innovation, in: *Policy Press* 7 (1/2022), 67–90.

O'Donovan, Cian/Smith, Adrian, Technology and Human Capabilities in UK Makerspaces, in: *Journal of Human Development and Capabilities* 21 (1/2020), 63–83.

Paris, Britt, The Internet of Futures Past: Values Trajectories of Networking Protocol Projects, in: *Science, Technology, & Human Values* 46 (5/2020), 1021–1047.

RoboCup Federation, Official Website, URL: https://www.robocup.org/ [last accessed: August 15, 2023].

Rommetveit, Kjetil/van Dijk, Niels/Gunnarsdóttir, Kristrún, Make Way for the Robots! Human- and Machine-Centricity in Constituting a European Public–Private Partnership, in: *Minerva* 58 (1/2020), 47–69.

Sadowski, Jathan/Bendor, Roy, Selling Smartness: Corporate Narratives and the Smart City as a Sociotechnical Imaginary, in: *Science, Technology, & Human Values* 44 (3/2019), 540–63.

Sadowski, Jathan/Pasquale, Frank, The Spectrum of Control: A Social Theory of the Smart City, in: *First Monday* 20 (7/2015).

Schneider, Sven/Hegger, Frederik/Hochgeschwender, Nico/Dwiputra, Rhama/Moriarty, Alexander/Berghofer, Jakob/Kraetzschmar, Gerhard K., Design and Development of a Benchmarking Testbed for the Factory of the Future, in: *2015 IEEE 20th Conference on Emerging Technologies Factory Automation (ETFA)*, Luxembourg 2015, 1–7.

Sclove, Richard E., *Democracy and Technology* (1st edition), New York 1995.

Smith, Adrian, Smart Cities Need Thick Data, Not Big Data, in: The Guardian, 18.04.2018, URL: http://www.theguardian.com/science/political-science/2018/apr/18/smart-cities-need-thick-data-not-big-data.

Smith, Adrian/Martín, Pedro P., Going Beyond the Smart City? Implementing Technopolitical Platforms for Urban Democracy in Madrid and Barcelona, in: *Journal of Urban Technology* (2020), 1–20.

Studley, Matthew E./Little, Hannah, Robots in Smart Cities, in: Maria I. Aldinhas Ferreira (ed.), *How Smart Is Your City? Technological Innovation, Ethics and Inclusiveness*, Cham 2021, 75–88.

Suchman, Lucy, *Human-Machine Reconfigurations: Plans and Situated Actions* (2nd edition), Cambridge, New York 2006.

Sydow, Jorg/Scheyogg, Georg/Koch, Jochen, Organizational Path Dependence: Opening the Black Box, in: *Academy of Management Review* 34 (4/2009), 1–21.

Tironi, Martín, Speculative Prototyping, Frictions and Counter-Participation: A Civic Intervention with Homeless Individuals, in: *Design Studies* 59 (2018), 117–38.

Tironi, Martín, Prototyping Public Friction: Exploring the Political Effects of Design Testing in Urban Space, in: *The British Journal of Sociology* 71 (3/2019), 1–17.

Valdez, Alan-Miguel/Cook, Matthew/Potter, Stephen, Roadmaps to Utopia: Tales of the Smart City, in: *Urban Studies* 55 (15/2018), 3385–3403.

While, Aidan H./Marvin, Simon/Kovacic, Mateja, Urban Robotic Experimentation: San Francisco, Tokyo and Dubai, in: *Urban Studies* 58 (4/2021), 769–86.

Zandbergen, Dorien/Uitermark, Justus, In Search of the Smart Citizen: Republican and Cybernetic Citizenship in the Smart City, in: *Urban Studies* 57 (8/2020), 1733–48.

Towards Placing Service Robots in Elderly Care Facilities

Rosalyn M. Langedijk, Kerstin Fischer

Abstract *This paper presents three studies about the development of robots for use in elderly care facilities and their employment in such settings. In Study 1, we aimed to uncover users' real needs by observing caregivers and residents in a Danish elderly care facility for 24 hours. The findings suggest that guidance robots might be useful and that drink-serving robots might not be as useful as initially anticipated in this particular facility. In Study 2, we tested the guiding task on one participant residing in the same facility. The results have revealed several opportunities for improvement regarding human–robot interaction and have provided insights on the difficulty of designing acceptable human–robot interactions. In Study 3, we visited a German elderly care facility, where we conducted a week-long field trial involving a drink-serving robot. Here, we faced several challenges regarding the actual deployment of robots in facilities. Finally, we present some ethical implications regarding consent forms, participants, and recordings of which researchers need to be aware when conducting real-world studies.*

1. Introduction[1]

Reducing caregivers' workload in care facilities is a key issue in elderly care. As Bodenhagen et al.[2] have noted, we are facing a societal challenge with the ongoing demographical changes, where, as life expectancy increases, so do overall health care expenses.[3] This presents a particular challenge for the everyday running of elderly

---

1 This project was supported by the Innovation Fund Denmark and carried out in the SMOOTH project framework (Seamless huMan–robot interactiOn fOr THe support of elderly people, Grant no.: 6158–00009B). We wish to thank the two elderly care facilities and Birgit Graf, Cagatay Odabasi, and Lotte Damsgaard Nissen, who helped during the field trial in Germany.

2 Leon Bodenhagen et al., Robot technology for future welfare: meeting upcoming societal challenges–an outlook with offset in the development in Scandinavia, in: *Health and Technology* 9 (3/2019b), 197–218.
3 Caroline Bähler et al., Multimorbidity, health care utilization and costs in an elderly community-dwelling population: a claims data based observational study, in: *BMC Health Services Research* 15 (1/2015), 1–12.

care facilities. One possible solution may be to have robots help with different kinds of tasks, for example, transportation, guidance, and serving beverages. However, little work has been done in relation to testing robotic solutions in real-world environments and gaining an understanding of what kinds of robotic solutions real users, such as caregivers and elderly persons, may need.

Field trials are an important step in that direction and in the development of human–robot interaction (HRI) research. The trials require the use of methods, such as ethnographic observations, that differ from those used in laboratory studies. By presenting results from three field studies, this paper sheds light on real users' needs as well as on the ways in which a robotic solution might support real-world tasks in elderly care facilities. Specifically, we address the following research question: What can we learn from field trials for robot development in elderly care facilities, and what methodological lessons can we obtain from the implications of field trials?

We report on three studies carried out within one project. The project's aim has been to create an interactional robot that supports staff and residents of elderly care facilities. The project developed a robot from scratch to address three use cases: serving fluids, collecting garbage and laundry, and guiding people. In this paper, we focus only on the interactional tasks, namely serving fluids and guiding people. Study 1 was a 24-hour observation of an elderly care facility in Denmark, where we collected data for all three use cases and gained an understanding of real users' needs. Study 2 was a field trial conducted in the same facility and involving a guiding robot. We conducted and analyzed a field test of the guidance scenario with a prototype of the final robot and a real user who was a resident of the elderly care facility. We did this together with roboticists, and we gathered data on both the technical and the interactional implications. Study 3 was a week-long field trial involving a drink-serving robot in an elderly care facility in Germany. For this, we collaborated with a German partner that contributed a specialized drink-serving robot.[4] Together, we collected information about the drink-serving use case and the implications of robots' possible employment for this purpose.

Overall, this paper presents observations, field trial results, and an analysis of the methodological and practical issues that arise when a robot is moved out of the laboratory. We sought to gain more knowledge about the use of robots in elderly care facilities and the kinds of impacts robots may have on residents and staff. Both test sites yielded relevant input on how researchers need to prepare to conduct real-world testing, and we shed light on different challenges that arise when testing in real-world settings.

---

4   Simon Baumgarten/Theo Jacobs/Birgit Graf, The robotic service assistant-relieving the nursing staff of workload, in: *ISR* 2018; 50th International Symposium on Robotics, Munich 2018, 1–4.

2. Ethnography and Field Trials in HRI

Generally, a field trial is conducted "in the wild," that is, in real-world environments and with real users. The aim of field trials is to determine how a product (in our case, a robot) works in an environment where it might be deployed in the future.[5] To conduct field trials in robotics, researchers need to find a facility where there are people who are willing to participate and where it would make sense to test a particular robot (in our case, one Danish facility and one German facility). Participants in field trials are mostly the real end-users; therefore, field trials elicit important feedback for robot designers on how their robot works within the given environment and how people react to it.

Generally, qualitative studies in HRI research are rare.[6] However, over the last couple of years, the number of qualitative studies, especially "in the wild" studies, has been slowly increasing. "In the wild" studies are conducted outside the lab, that is, in real-world environments (e.g. field trials). In this area, ethnography is a commonly used method that enables researchers to gain insights that other methods do not facilitate, for example, to understand how robots could fit into a workflow[7] or to understand the impact and usage of robot technology in homes.[8] However, Hasse et al. have pointed out that ethnography is still "more or less absent in the field"[9] and that HRI researchers use ethnography quite differently. According to her, ethnography offers much more than just anecdotes, and ethnographic researchers do more than "just looking."[10] She has also reminded HRI researchers that ethnography carries out research with people and not about people, which means that researchers include people, their feedback, and the overall context in their research. Thus, ethnog-

---

5   Cf. Selma Sabanovic/Marek P. Michalowski/Reid Simmons, Robots in the wild: Observing human-robot social interaction outside the lab, in: *9th IEEE International Workshop on Advanced Motion Control*, Istanbul 2006, 596–601.

6   Louise Veling/Conor McGinn, Qualitative Research in HRI: A Review and Taxonomy, in: *International Journal of Social Robotics* 13 (2021), 1–21.

7   E.g. Bilge Mutlu/Jodi Forlizzi, Robots in organizations: the role of workflow, social, and environmental factors in human-robot interaction, in: *HRI '08: Proceedings of the 3rd ACM/IEEE International Conference on Human-Robot Interaction* 2 (2020), 287–294.

8   E.g. Jodi Forlizzi, How robotic products become social products: an ethnographic study of cleaning in the home, in: *HRI'07: Proceedings of the ACM/IEEE International Conference on Human-Robot Interaction* 2007, 129–136; Julia Fink et al., Living with a vacuum cleaning robot, in: *International Journal of Social Robotics* 5 (3/2013), 389–408.

9   Cathrine Hasse/Stine Trentemøller/Jessica Sorenson, Special issue on ethnography in human-robot interaction research, in: *Journal of Behavioral Robotics* 10 (1/2019), 180–181.

10   Hasse/Trentemøller/Sorenson, Special issue on ethnography in human-robot interaction research.

raphy offers rich data and provides information for future developments and design processes.[11]

Ethnographic fieldwork is increasingly being used in HRI research to test robots in real-world environments with real users, that is, "in the wild."[12] "In the wild" testing is complex because it involves unpredictable events and untrained users.[13] In comparison, in lab studies, everything is controlled for, dialogues are often scripted,[14] and the robot is often controlled by wizards (the people who operate the robot). The reason lab studies dominate the field of HRI is the focus on replicability, representativity, and reliability,[15] which are much harder to obtain in real-world scenarios with diverse users who have their own agendas, preferences, and schedules in an uncontrolled situation where unforeseen obstacles may occur, including unplanned participants and overhearers; such situations cannot be controlled so as to exclude other contingent factors.[16] These realities are understood as rendering HRI experiments unreliable and not replicable, characteristics that lack adherence to the high scientific standards. In contrast, Jung and Hinds have argued that it is time to take robots into the real world since controlled lab studies cannot replace field trials.[17] Similarly, Sabanovic et al. have stated that researchers need to do real-world testing to reveal the real potential of robotics solutions because

---

11   Ylva Fernaeus et al., How do you play with a robotic toy animal? A long-term study of Pleo, in: *IDC'10: Proceedings of the 9th international Conference on interaction Design and Children* 2010, 39–48; Pericle Salvini/Cecilia Laschi/Paolo Dario, Design for acceptability: improving robots' coexistence in human society, in: *International journal of social robotics* 2 (4/2010), 451–460; JaYoung Sung/ Henrik I. Christensen/ Rebecca E. Grinter, Robots in the wild: understanding long-term use, in: *Proceedings of the 4th ACM/IEEE international conference on Human robot interaction*, 2009, 45–52.)

12   Wang-Ling Chang/Selma Šabanović/Lesa Huber, Situated analysis of interactions between cognitively impaired older adults and the therapeutic robot PARO, in: *International Conference on Social Robotics* 2013, 371–380; Forlizzi, How robotic products become social products; Sung et al., Robots in the wild; Alessandra M. Sabelli/ Takayuki Kanda/ Norihiro Hagita, A conversational robot in an elderly care center: an ethnographic study, in: *6th ACM/IEEE international conference on human-robot interaction*, 2011, 37–44.)

13   Antonio Andriella/Carme Torras/Guillem Alenyà, Short-term human–robot interaction adaptability in real-world environments, in: *International Journal of Social Robotics* 12 (2020), 639–57.

14   Cf. Guy Hoffman/Xuan Zhao, A primer for conducting experiments in human–robot interaction, in: *ACM Transactions on Human-Robot Interaction* 10 (1/2020), 1–31.

15   Cf. Hee R. Lee et al., Configuring Humans: What Roles Humans Play in HRI Research, in: *Proceedings of the IEEE Human-Robot Interaction Conference*, Sapporo 2022.

16   Cf. Hoffmann/Zhao, A primer for conducting experiments in human–robot interaction.

17   Malte Jung/Pamela Hinds, Robots in the wild: A time for more robust theories of human-robot interaction, in: *ACM Trans. Hum.-Robot Interact.* 7 (1/2018), Article 2.

how stakeholders in actual situations interact with a robot cannot be ascertained through lab studies.[18]

Detailed descriptions of current workflows and people's unstaged interactions and real-life conduct can be studied using ethnography.[19] Instead of testing pre-specified hypotheses, ethnography enables the researcher to explore the nature of the context[20] and describe the observed practices from people's own perspectives.[21]

Ethnography is a qualitative method that generates a holistic understanding of communities of practice, and it is particularly useful in providing "the native's point of view."[22] According to Blomberg et al., ethnographic practice has four basic principles: a natural setting, a holistic perspective, descriptive understanding, and members' point(s) of view.[23]

The requirement that the research has to take place in a natural setting means that the researcher has to enter the field of interest, where people will do as they always do: engage in unstaged activities. Typically, HRI research takes place in a lab;[24] ethnography, in contrast, allows the researcher to carry out research outside the lab. Nevertheless, there may be different degrees of "staging" of the interactions (cf. our discussion of Study 2 in Section 5.3).

In an ethnographic approach, activities in natural settings are studied holistically, which means that we observe anything as relevant and every aspect as being as important as the other—without interpreting the actual observations. Instead of focusing on controlled one-on-one interactions between humans and robots, an ethnographic approach also takes the context into account, as well as the workflow in which the robot's actions will be integrated, a large range of stakeholders beyond the robot's direct "user", and the whole organizational and institutional context in which the interaction is embedded.

Furthermore, ethnography is a type of fieldwork in which the researcher describes the activities from the members' viewpoint(s), aiming to uncover the group members' tacit knowledge. In our studies, the setting cannot be described as "natural" because the robot was either wizarded for security reasons (Study 3) or was under close surveillance by a research team that was also filming the interactions (Study 2), or the researchers had to facilitate the interactions (Studies 2 and 3). Thus,

---

18   Sabanovic/Michalowski/Simmons, Robots in the wild.

19   E.g. Leon Bodenhagen et al., Robot use cases for real needs: A large-scale ethnographic case study, in: *Journal of Behavioral Robotics* 10 (1/2019a), 193–206.

20   Scott Reeves/Ayelet Kuper/Brian D. Hodges, *Qualitative research methodologies: ethnography*. BMJ 2008, 337.

21   Jeanette Blomberg/Mark Burrell/G. Guest, An ethnographic approach to design, in: *The Human-Computer Interaction Handbook*. L. L: Erlbaum Associates Inc., Hillsdale 2003.

22   Martyn Hammersley/Paul Atkinson, *Ethnography: Principles in practice*, New York 2007.

23   Blomberg et al., An ethnographic approach to design.

24   Cf. Hoffman/Zhao, A primer for conducting experiments in human–robot interaction.

the ethnographic analysis reported in this paper has some experimental aspects that need to be considered in the evaluation of the case studies' results. The reason researchers cannot currently carry out real ethnographic tests with robots is because the robots are still at the prototype-level, which does not allow for realistic robot deployments.

The methods we used for our field trials were mostly the same as those used in other ethnographic studies and so were the data elicited, namely observation notes, recordings, and in-depth interviews with users, which are qualitative data. Regarding observations, researchers observe what is available: what people do, what they say, and how they work (from their own perspective as members of a specific group). The results comprise a broad range of observation notes, artifacts, pictures, sketches, and so on. Because it is impossible to observe everything, many scholars suggest having a framework to record observations—and having a focus.[25] There are many frameworks for observations,[26] and because we were interested in how activities take place today and the requirements to have a robot carry them out, we created the following observation scheme: activities (and how the activities take place), participants (who is involved in the observed activity), time (when and for how long the activity takes place), objects (which objects are involved, e.g., the type of cup or cutlery), and environment (where the activity takes place and whether anything special happens around the activity). This scheme included the most important aspects that we needed to pay attention to in order to gain an understanding of everyday life in elderly care facilities as well as how a robot could be employed in those settings. The scheme was printed in A5 booklets with a scheme on the right side and a blank page on the left and followed by two pages for notes. Having a scheme helps observers structure their observations and focus on the relevant aspects.

Theoretically, an observation study can be concluded when no new topics emerge and the observations become repetitive. However, in reality, practical and/or financial issues often determine the length of an observation study; in our case, a 24-hour observation (in the Danish institution) and a week-long field test (in the German institution) had previously been negotiated with the respective facilities.

An important distinction can be made concerning the degree of the researcher's participation.[27] It varies across a spectrum from insider to outsider, or, as in the ex-

25   E.g. Natasha Mack et al. *Qualitative research methods: A data collector's field guide*, North Carolina 2005.

26   E.g. Vijay Kumar/ Patrick Whitney, Faster, cheaper, deeper user research, in: *Design Management Journal (Former Series)* 14 (2/2003), 50–57; James P. Spradley, *Participant observation*, New York 1980; Christina Wasson, Ethnography in the field of design, in: *Human organization* 59 (4/2000), 377–388.

27   Y. Rogers/H. Sharp/J. Preece, *Interaction design: beyond human-computer interaction* (4th edition). New Jersey 2015.

ample Blomberg et al. have described, from participant-observer to observer-participant.[28] As an outsider, the researcher is a "fly on the wall" that observes from outside the group, which means that they do not engage with the group. As an insider, the researcher becomes "one of them"; that is, they become a member of the group and engage as such. The observer's role is to behave so that they do not affect the natural flow of the activity.[29] In our Studies 1 and 3, however, we experienced some difficulties with this distinction (see our discussion below).

In the current paper, we report on using ethnography as a research method to obtain insights into people's (in our case, caregivers' and residents') everyday caregiving-related practices to discover complex real-world practices. We did this through ethnographic observations and in-depth interviews, which yielded an understanding of relationships, processes, and expectations, rather than a final product.[30] Furthermore, we used ethnographic fieldwork as a method to test how a robot would work in an elderly care facility, either as a guiding robot or as a drink-serving robot. However, "in the wild" studies bring specific challenges, such as collecting consent forms.[31] We will discuss these issues in Section 7.

3. Case Studies

In this section, we present the three case studies and elucidate how they relate to one another. Based on previous discussions with the facility management, two interactional routine tasks were deemed interesting to focus on during the observations, namely providing walking guidance and serving drinks. The methodology employed combined participant observations and field trials since we observed both one-on-one interactions and the overall contexts in which the interactions took place.

The participant observations comprised observation and in-depth interviews with residents and care personnel. The field trials comprised interactions with and around the robot, including among the researchers, residents, and care personnel. The studies revealed numerous practical findings, which led to robot design recommendations and the identification of situations in which a robot could be beneficially deployed.

---

28   Blomberg/Burell/Guest, An ethnographic approach to design.

29   Mack, Qualitative research methods.

30   Bohkyung Chun/Heather Knight, The Robot Makers: An Ethnography of Anthropomorphism at a Robotics Company, in: *ACM Transactions on Human-Robot Interaction* 9 (3/2020), 1–36.

31   Cf. Rosalyn M. Langedijk et al., Studying Drink-Serving Service Robots in the Real World, in: *2020 29th IEEE International Conference on Robot and Human Interactive Communication (RO-MAN)*, Naples 2020, 788–793.

3.1   Study 1

Study 1 comprised 24 hours of observations in a Danish elderly care facility. We observed workflows and care practices in interactions between caregivers and residents. We argue that understanding real users' needs is important to design helpful robotic solutions. The first step is to understand whether there really is a need. Next, it is necessary to understand whether and how a robotic solution might support this need. Study 1's findings provided us with initial insights into workflows, procedures, and opportunities for robotic support. Thus, we named this study "Understanding Needs."

3.2   Study 2

Study 2 was based on our observations during Study 1. Here, we observed that walking guidance may be a helpful task for a robot to execute. This study was carried out in the Danish facility. The robot guided one participant, and we focused on the micro-sociological aspects of the interaction. Consequently, we argue that a detailed understanding of processes in interaction in general is crucial for testing in real-world environments. Thus, we named this study "Understanding Interaction."

3.3   Study 3

Study 3 aimed to provide an understanding of the issues that arise when employing robots in elderly care institutions. During our observations in Study 1, we saw that a drink-serving robot would not make sense in that particular care facility. However, we found a different facility where the management agreed that such a robot could indeed be helpful. We used a different robot designed to serve a variety of drinks and conducted a week-long field trial in which the robot served various types of beverages in a German elderly care facility. The findings provided us with feedback on the challenges that arise methodologically as well as concerning the deployment of robots in such a context. Thus, we named this study "Understanding the Employment of Robots in Institutions."

4. Study 1: Understanding Needs

Study 1 was carried out as part of a needs analysis to understand caregivers' and residents' everyday practices and how a robot might support them.

4.1   Procedure

To conduct this study, we visited a Danish elderly care facility for 24 hours and collected observational data, but due to data protection legislation and the health status of the residents, which did not allow them to provide informed consent, we were only permitted to take observation notes; audio and video recording were prohibited. The researcher (first author) took notes in an A5 booklet and largely shadowed one caregiver.

Initially, we planned a "fly on the wall" approach, but this was not feasible since the caregivers and residents engaged with the researcher quite often and included her in their conversations. Therefore, the researcher spontaneously decided that it was best to change the approach and become "one of them," which entailed eating meals with them and interacting with them as if the researcher was part of the facility.

During the observations, we focused on the daily routine tasks that caregivers fulfill and how they communicate with the residents. Robots may be capable of performing daily routine tasks, and our goal was to evaluate whether this was feasible. Because those were the tasks deemed to be the most feasible technically, our observations mainly focused on the activities "providing guidance" and "serving beverages." We observed who participated in the activities, when they took place, their duration, and the environment in which they occurred (see the coding scheme). We also observed verbal interactions to inform robot dialogue. We observed how caregivers got residents' attention, what they said to residents and how they said it, and finally, how residents reacted to caregivers, both verbally and nonverbally.

In addition to the observations, we had the opportunity to conduct in-depth interviews with the caregivers at the facility during quiet times in the afternoon and evening. We did not want to interrupt or disturb the workflow; therefore, we collected our questions and observations to be addressed during the in-depth interviews with the care personnel.[32] These interviews were mainly for clarification of the background informing certain practices, but we also covered how well the interviewees liked the robotic solutions we presented to them.

4.2   The Facility and the Participants

The Danish elderly care facility has four small units, each of which houses five to six residents. The two observed units differed considerably from each other, mainly because the residents themselves were very different. In the so-called "pink" unit, there were a total of five residents, one in a wheelchair, two using walkers, and two who were ambulatory without assistance. Those residents were physically fitter, and

---

32   Cf. Blomberg/Burell/Guest, An ethnographic approach to design.

there was much more noise in their unit than in the "blue" unit, which had six residents. Of those, two were in a wheelchair, one used an electric wheelchair, one used a walker, and two did not require any assistance.

Not all residents were study participants. Many of the residents suffered from dementia, so very few (1–2 residents per unit) were able to interact with the researcher. Participants' informed consent was collected by representatives of the institution prior to our visit.

4.3   Findings

Our findings concern the provision of guidance for residents when moving between their rooms and the dining area, as well as drink-serving interactions. These were identified beforehand as potential areas in which a robot might usefully support caregivers in their work.[33]

4.3.1 The Guiding Task

Caregivers spend a large amount of time guiding residents from points A to B, for example, from their rooms to the dining area to have a meal. We observed that even when the residents said that they were not going, they often forgot saying that within the next minute. We also saw that some residents got lost and sat down in confusion when the caregiver left them to attend to another resident.

In one instance, at lunchtime, a caregiver knocked on four doors in a row and announced, "Lunch is ready." Although the roused residents could walk by themselves in principle, some still needed guidance. In this situation, the third resident who had been called looked around desperately when summoned, not knowing what to do. When the caregiver exited the fourth room, she instructed the resident to go to the dining area. However, the resident was confused and instead sat on a nearby chair. The fourth resident tried to motivate him to come along, but he preferred to wait for the caregiver to exit a sixth resident's room, accompanied by a resident in a wheelchair. When the caregiver approached, she verbally motivated him to stand up and start walking, which he eventually did. This instance illustrates individual residents' differing needs, as well as the importance of repeated motivational cues from a guide in these types of situations. Similar situations involving different residents were observed before every mealtime.

During our interviews, the caregivers mentioned that they could see the relevance of a robot helping with this task, especially in the form of a "travelling companion"; for instance, a robotic "guide" could be customized to evoke personal memories. Employed in this role, the robot could store a large amount of information

---

33   Cf. Kerstin Fischer et al., Integrative Social Robotics Hands-On, in: *Interaction Studies* 21 (1/2020), 145–191.

and present it in multimodal ways, thereby carrying out a task that humans should but cannot do.[34] In so doing, the robot would provide even more added value than human caregivers.

A robotic solution would also be useful when residents (that is, all who wish to) join one of the group activities held outside each unit or one of the activities held in the common area shared by all four units. The robot could guide residents from their respective units and back again. Some residents cannot go by themselves and need support, whereas others cannot manage to find their way back to their rooms afterwards. However, no group activities were held during our stay, and we could not observe how the residents and caregivers perform the task; we were therefore limited to caregivers' descriptions of how it is normally done.

We also had the opportunity to discuss this task with one of the residents. He liked the idea of a companion very much. Although he could still walk independently with the use of a walker at the time of the conversation, he often forgot where he needed to go. Nevertheless, he stressed his preference for a companion and not a guide.

In sum, the guiding activity is mainly carried out with one caregiver and one resident; however, in some instances, other residents try to motivate a resident to continue to walk as well. Guidance takes place within each unit, often around meal-times. Based on our observations, we believe that guiding could be a useful task for a robot to perform in an elderly care facility. We did not anticipate these patterns when we entered the facility to observe the caregivers' and residents' everyday life challenges.

4.3.2 The Drink-Serving Task

Another possible use for a robot could be to drive around offering people beverages given that older persons tend to lose their sense of thirst and thus need to be monitored and reminded to drink sufficient fluids. We observed five meals during which we saw that the different types of fluids served varied considerably and were highly personalized: Residents were provided with red or yellow lemonade (with or without sugar) and with many different varieties of coffee (e.g., black coffee, coffee with cream or milk, coffee with sugar, coffee with cream and sugar, etc.). We also observed highly personalized containers, such as glasses, plastic cups, and cups with straws.

Additionally, we found that many restrictions apply; for example, residents may not choose their own beverages, may not remember what they should and should not have, or may not be able to physically take their own beverage. For instance, it could be dangerous for a person with diabetes to consume a glass of lemonade with sugar.

---

34   Cf. Johanna Seibt/Malene F. Damholdt/Christina Vestergaard, Integrative social robotics, value-driven design, and transdisciplinarity, in: *Interaction Studies* 21 (1/2020), 111–144.

For this reason, the caregivers indicated that it is easier if they serve the drinks themselves instead of ensuring that individual residents take the right beverage from a robot. This is a highly practical implication for a robotic solution, and we had no prior knowledge of it.

Regarding motivating residents to drink sufficient fluids, we observed one instance during breakfast when a resident did not want to drink her lemonade. The caregiver tried to encourage her to drink several times, and we observed that an effective method during meals is to say, "Skål," which invokes a common Danish toasting ritual. When someone says, "Skål," it is customary for everyone to toast and drink. Returning to the example, the caregiver constantly reminded the resident to take a sip. Finally, the caregiver sat next to her and told her that she needed the glass to turn on the dishwasher. The resident could not hear that the dishwasher was already on, and the fib worked; the resident took the last sip, but she sat at the table in her wheelchair for 45 minutes before she finally finished her lemonade. Afterwards, the caregiver told us that it is caregivers' practice to insist in the manner demonstrated because that resident does not drink enough. This example shows the importance of repeated motivational cues and patience and thus the challenge of implementing a robotic solution for this type of task.

In sum, the activity involves a caregiver trying to motivate a resident to drink or a single caregiver serving all the residents a beverage, for example, with a meal. We observed the activity in the common area, usually around mealtimes. Specifically, we noted that many different objects were involved, including the different containers from which residents drank. Based on our observations, we concluded that it is not possible to implement a drink-serving robot in this elderly care facility for several reasons. First, the residents do not drink water at all, nor do they drink any other single type of beverage. They all have very different preferences, as well as certain beverages they cannot have due to diseases such as diabetes. Second, the residents use different types of cups, for example, glasses, plastic cups, or cups with a straw, depending on which container they are able to use. Third, residents may not be able to retrieve a beverage from a tray themselves; the beverage needs to be placed in front of them. Nevertheless, encouraging residents to drink enough was verified as a necessary but tiresome task for caregivers.

4.4   Discussion

The results of this study show that a robot conducting the guiding task might be useful in an elderly care facility. With this finding, we ran a field trial in Study 2 to test the feasibility of implementing a guiding robot. Furthermore, our in-depth interviews have revealed that both residents and caregivers make the important linguistic distinction between a "companion" and a "guide." The latter reminds the elderly about things they cannot do anymore, whereas a "companion" is there to help and provide companionship. The word "companion" is associated with a much kinder image. This information is important when researchers and robot designers introduce a robot to a new audience since the first impression based on the initial presentation of the robot may influence people's willingness to interact with it in the future.

Furthermore, the results show the unfeasibility of employing a robot to perform the drink-serving task in this particular care facility. Because of the high variability in both drinks and drinking containers, a robotic solution would be hard to implement. However, we ran a field trial in Study 3 in a different care facility, where the management saw great potential for a robotic solution for this routine task.

In our observations, we noticed the use of many motivational cues in both tasks as well as in numerous different scenarios, indicating the utility of a robot that can motivate its interaction partners. Specifically, the robot could motivate residents to begin an activity, such as going to lunch or having a beverage; continue with an activity, such as walking; or restart a prior activity, such as resume walking after stopping or finish the contents of their glass.

Regarding the research methodology, the observation approach was changed from "fly on the wall" (passive) observation to active participant observation. If we had continued with passive observation, the residents and caregivers would have felt highly scrutinized and might have changed their activities due to the researcher's presence. Becoming a moderate participant made it easier to get to know the caregivers and residents and their everyday work practices. Furthermore, in several instances, it was not possible to remain an outsider, for example, when a resident asked for help or addressed the researcher personally. Additionally, at the beginning of the observations, a caregiver who was being observed provided profuse explanations and tried to justify her actions, which showed us that she actually felt observed. However, this behavior diminished over time as the caregiver familiarized with the situation.

In sum, our observations show that a guiding robot may be useful. Therefore, in Study 2, we conducted a field trial with a guiding robot at the same facility. Our observations also show that the envisioned drink-serving robot is not feasible in this specific facility. However, a robot that can serve a variety of drinks might be useful in a different facility. Thus, in Study 3, we conducted a field trial with a different robot in a German elderly care facility to test whether a drink-serving robot would be feasible there.

5. Study 2: Understanding Interaction

5.1   Procedure

To conduct this field trial, we returned to the Danish elderly care facility (the same as in Study 1) with the SMOOTH robot[35] (see Fig. 1) and conducted initial tests on the guiding activity. The aim of the trial was to guide a resident from one point (preferably their room) to the dining area. However, the guidance started in the middle of the hallway due to hygiene regulations that disallowed the robot from entering residents' private rooms. We tested the robot's autonomous navigation and dialogue capabilities.

At the beginning of the test, a resident, that is, an older person in need of guidance, was seated in a chair in the middle of the hallway. The robot picked up the resident and walked in front of him, while adapting its speed to the resident's pace and providing motivating speech to guide him to the dining area.

The situation was staged in several ways. First, for safety reasons, the robot's behavior was closely monitored by the engineering team, and an additional person videotaped the interaction. Furthermore, the caregiver and the resident were situated in a certain location, waiting for the robot to arrive; the location likely would not be the real starting point for the guidance. Additionally, the resident knew that the aim of the interaction was to test the robot, and thus, he had no intrinsic motivation to get to the dining area in that moment. From that perspective, the field test results do not shed light on the future situations in which robots can be used. On the other hand, the test allowed for the micro-sociological analysis of aspects of HRI and has provided us with an idea of what kinds of capabilities a robot employed in an elderly care facility in the future would need to have. Furthermore, the environment was familiar and authentic to the resident, unlike a lab environment.

5.1.1   The Robot

The SMOOTH robot[36] is a large service robot developed to take over several tasks in elderly care facilities. The robot was designed as a modular mobile robot platform, which ensures that it can easily be prepared for different tasks. The robot's head has a microphone, speakers, cameras, and two touchscreens, one in the front and one in the back. The front touchscreen displays simulated eyes. Furthermore, the robot is equipped with autonomous navigation and dialogue capabilities.

---

35   Norbert Krüger et al., The SMOOTH-robot: a modular, interactive service robot, in: *Frontiers in Robotics and AI* 8 (2021).

36   William K. Juel et al., Smooth robot: Design for a novel modular welfare robot, in: *Journal of Intelligent & Robotic Systems* 98 (1/2020), 19–37.

Fig. 1: SMOOTH robot

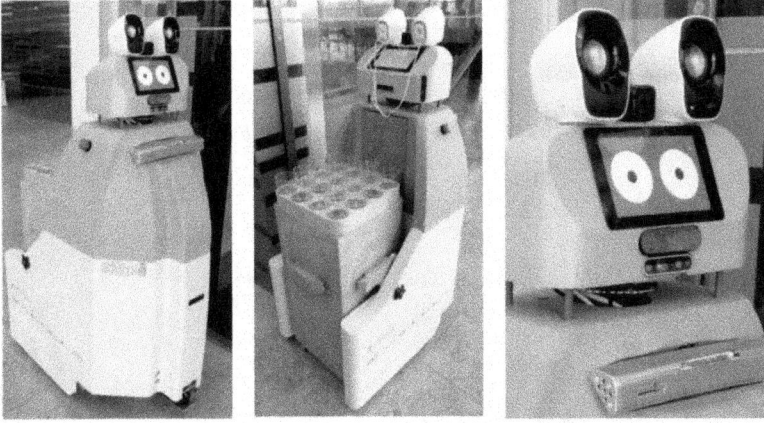

5.1.2 The Participants

In this study, we only had access to one resident due to the other residents' health condition at the time. The participant was an older man who used a walker to move around the facility. Other participants were researchers, who assisted the resident, observed the scenario and the robot, and took photos and recorded the interactions. None of the other residents or staff were in the area during our tests. Since only researchers, staff, and the resident who consented to participate in our project were present, we videotaped the interaction.

5.2   Findings

We will describe one interaction as an example for linkage with the observations (see the YouTube video).[37] The video analysis allowed us to watch the interaction repeatedly and analyze it in greater detail than would have been possible if we had been limited to observation notes. We also checked our observation notes for redundancy and found that combining both types of data gave us a more detailed analysis.

---

37   https://www.youtube.com/watch?v=AbK_83Qy6do [last accessed: August 15, 2023].

1) First, a researcher pressed a button on the robot to start the guiding activity. The robot moved to the start position, which, in this case, was in the middle of the hallway, near the resident.

2) Next, the caregiver spoke to the robot (see Fig. 2): "Hi, SMOOTH." This was necessary because the elderly resident could not speak loudly enough for the speech recognition system to recognize that he was summoning the robot.

Fig. 2: The robot arrives

3) The robot acknowledged that it was being summoned by saying, "Yes, I am coming," and added, "I am on my way." This was followed by a long silence while the robot approached.

4) When the robot approached, it established eye contact with the resident. The assisting researcher instructed the robot to guide the resident to the dining area (see Fig. 3): "SMOOTH, guide Poul to the dining area." The robot acknowledged the request by saying, "Alright, just follow me."

Fig. 3: The robot establishes mutual gaze

Smooth recognises Poul as resident so it approaches

5) Next, the resident stood up and started following the robot, while the robot turned to start the guided journey. While they are walking, there was a large distance between the older person and the robot (see Fig. 4) that persisted throughout the guided journey.

6) The robot stopped suddenly to prepare to take a turn, which prompted the older person to monitor the robot closely.

Fig. 4: The robot guides the resident

7) The elderly resident had obvious difficulty making a smooth turn with his walker, given that the robot had stopped and turned very sharply (see Fig. 5).

Fig. 5: The robot turns and provides motivational speech

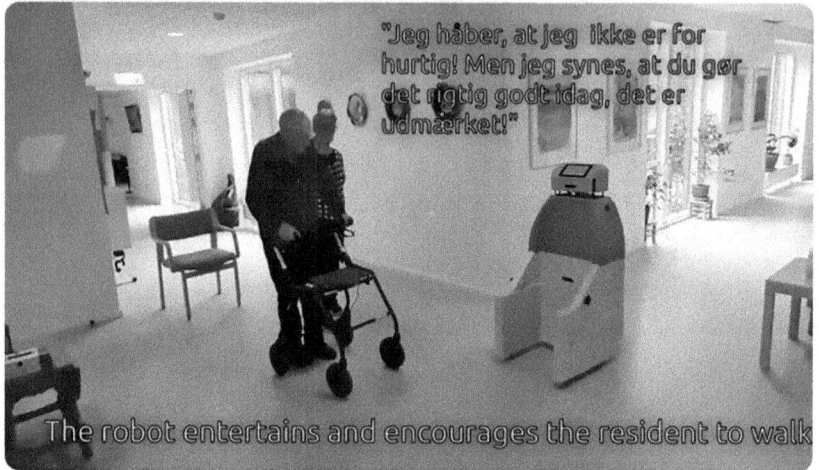

8) Regarding dialogue, the robot directed its speech in the direction opposite the resident's location, and the large distance between them made it even harder for the older person to understand that the robot was talking to him. As the interview we conducted at the end of the session revealed, the resident did not even realize that the robot's output had been directed at him, let alone understand what it said. Nevertheless, the robot produced motivational cues, for instance, "You are doing great today," and also engaged in small talk about dinner and local news, for example, by saying, "The local handball club is doing great this season." (see Fig. 6)

9) When they arrived at their destination, the robot said, "Now we are here. Have a nice day." The robot then turned away from the destination and stood in the middle of the room (see Fig. 7). The older person looked around, wondering what to do, as he did not hear the robot say that they had arrived at their destination.

Fig. 6: The resident is following the robot in some distance

Fig. 7: The robot arrives

10) Eventually, the elderly person being guided walked past the robot, and a care-giver took over to direct him to his seat (see Fig. 8).

Fig. 8: The resident needs to walk around the robot

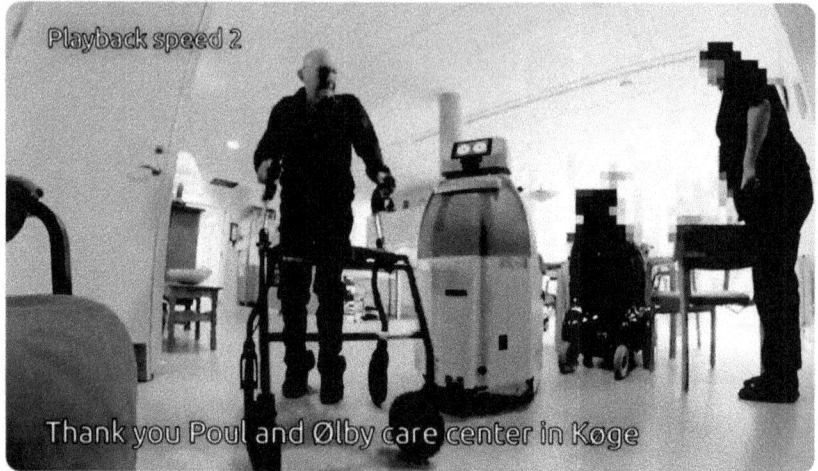

5.3   Discussion

The field trial has given us insights into real-world testing and the feasibility of using the SMOOTH robot to provide navigation guidance. We will discuss several observations and share recommendations for the dialogue and navigation functions.

First, from a technical perspective, the field trial was successful because the robot adapted its speed seamlessly to that of the resident and kept the same distance throughout the guided journey.[38] However, there were general problems with dialogue and speech recognition. To start the guidance, an assisting researcher had to call the robot because its speech recognition did not recognize the older person's speech. This is a crucial point because it means that a caregiver would need to be present to start the guidance each time the robot is used for the guiding task. Next, when the robot approached the resident, it said, "I am coming," and shortly after, it said, "I am on my way." The idea underlying these two utterances was that they could fill the silence while the robot approached someone. However, the sentences were uttered in quick succession, leaving several seconds during which nothing was said, thus creating a period of time when it was unclear what the robot was doing. During that time, the elderly resident just sat and waited for the robot to arrive. On

---

38   See Krüger et al., The SMOOTH-robot.

this, we referred to Fischer et al., who found that participants perceived silences such as these as uncomfortable.[39]

We also noted several other problems regarding unpredictable robot behavior. For example, the robot stopped suddenly to turn a corner, which made the older person who was following it uneasy, as evidenced by his subsequent close monitoring of the robot. Again, based on our findings in reference to Fischer et al., we have assumed that if the elder had been able to hear the robot, he might have felt more comfortable, and it would have been easier for him to anticipate the next action.[40] Moreover, the elder had to struggle because the robot did not anticipate the difficulty he would experience when turning the corner with his walker. To address this, we recommend that the robot slow down when making a turn instead of stopping and turning abruptly.

The large distance between the elder and the robot was also problematic. When the guided journey started, we observed a significant distance between the older person and the robot that appeared to be rather unnatural in a guiding scenario. Being a considerable distance apart makes it hard for the human and the robot to interact with each other, which is problematic regarding opportunities for small talk.

Furthermore, the interview we conducted after the interaction has revealed that the older adult had problems understanding the robot's speech. Since the robot directed its speech in the opposite direction, the elder did not know that the robot was talking to him at all. If the robot had been able to turn its head, the elder might have become aware of the attempted interaction. Ideally, the guidance would be provided with the robot and the human situated side by side. In that case, the distance would be much smaller, and the person being guided should have little difficulty hearing the robot's utterances. The close proximity would have the added benefit of allowing the person being guided to pay less attention to the robot's movements because the possibility of an elder bumping into the robot when it stops to turn, for example, would be greatly reduced. Additionally, in the field trial, because the elder being guided did not hear the robot's speech, he did not know when the guided journey had ended and was unsure why the robot had stopped in the middle of the hallway near the dining area.[41]

This leads us to the last observation about the robot's final positioning at the destination. Currently, the robot's end position is inconvenient because caregivers and residents need the space to navigate. Furthermore, the caregiver was needed to take over and guide the resident to his seat. Possibilities for improvement in this

---

39 Kerstin Fischer/Hanna M. Weigelin/Leon Bodenhagen, Leon, Increasing trust in human–robot medical interactions: effects of transparency and adaptability, in: *Journal of Behavioral Robotics* 9 (1/2018), 95–109.

40 Fischer/Weigelin/Bodenhagen, Increasing trust in human–robot medical interactions.

41 See also Fischer et al., Integrative Social Robotics Hands-On, for a discussion of this issue.

area include the robot guiding the person to their seat at the table, which is the real destination (not just the dining area in general), or the robot moving out of the way, presupposing that it can anticipate the user's next action and movement trajectory.

Concerning the methodological implications, the test's setting was not naturally occurring; instead, to facilitate human-robot interactions, the field was essentially turned into a lab.[42] Furthermore, the setting was staged, as a chair was placed in the middle of the hallway. Finally, the testing was conducted with only one participant, as only one resident of the chosen facility was healthy enough to participate. The very small sample size raises the question of whether he was a good representative of the facility's overall elder population. However, the field trial was useful because it provided initial insights into the interactional aspects of navigation guidance. During the overall project, we found significant differences in the mobility of residents across various elderly care institutions; therefore, it is likely that there will be residents in other facilities for whom the navigation guidance scenario the robot can offer may be very useful indeed.[43] Moreover, the field trial was positively perceived by the resident and the care personnel, and thus, we can conclude that the guiding task is feasible in principle. However, it should be further improved and retested, especially in different environments and with more people.

6. Study 3: Understanding the Employment of Robots in Institutions

Study 3 was a field trial conducted in a German elderly care facility, involving the Robotic Service Assistant (see Fig. 9), namely a robot specifically developed to serve different kinds of beverages. The robot was operated in a common area, where it served water, apple juice, and orange juice during the day.

6.1   Procedure

We contacted the care facility and established a mutual goal. We explained what the robot was capable of doing and what we wanted to investigate, and the management asked us what they wanted to know. The management was excited about the collaboration and looked forward to the staff's and residents' reactions as well as to our results. We also discussed which drinks to serve because Study 1 had taught us that the kinds of drinks served could be restricted. Since water is always available in the German facility, we decided to have the robot offer other drinks (in addition

---

42   Cf. Lee et al., Configuring Humans.

43   Cf. Denise Hebesberger, What do staff in eldercare want a robot for? An assessment of potential tasks and user requirements for a long-term deployment, in: IROS Workshop on "Bridging user needs to deployed applications of service robots, Hamburg 2015.

to water) that the residents do not usually get, namely apple juice and orange juice, to increase the robot's attractiveness in their eyes. We did this because we aimed to record as many interactions as possible and could not anticipate how the residents would react to the robot. For that reason, we wanted to offer something tangible that would be attractive to them; however, it turned out that the residents who signed up for the study were eager to interact with the robot mainly because they found it interesting.

Next, we discussed how the management could prepare the staff and the residents and their relatives for the testing and what was needed before we could start. Much time was spent preparing consent forms and following the procedures for obtaining consent. We visited the facility to speak with management in person, present the consent forms, and familiarize the staff with the procedure for collecting consent forms; the staff then took over the task of collecting the residents' consent paperwork.

A researcher (first author) and one colleague introduced the robot and the project itself to the residents by visiting the units, as well as by talking to the residents and caregivers while our engineering colleagues were mapping the area to prepare the setting. The robot was set up in a common area, namely one of the two dining areas.

The robot can navigate autonomously; however, for safety reasons we used a wizard to control the navigation. This wizard stood 1.5 m behind the robot and controlled it with a joystick. We used another wizard to manage the dialogue. That wizard sat in the same room as the robot, so that she could hear what was being said. Neither of the two wizards were hidden from the participants' view because of the structure of the environment. Furthermore, one researcher functioned as an observer and was available to the residents in case they needed help. A fourth researcher joined us for two days and conducted technical tests when there were no residents present. This setting was as natural as possible since the robot could not operate autonomously due to safety issues. The robot addressed residents who had consented to participate in our study among those sitting in the common area at the tables. Additionally, the staff suggested taking the robot to common activities, where the robot offered drinks to everyone who wanted one (irrespective of whether they had completed a consent form).

6.1.1   The Facility and the Participants

The elderly care facility is a large five-floor building, and each floor is a separate unit that houses 15–25 residents and approximately three to four caregivers. On the fourth floor, where we carried out our study, there were 24 residents (aged between 60 years and 94 years). We collected consent forms from all 24 residents. Six of the residents were absent for the entire week during which the study was conducted, but another 6 of the 24 residents were in the common area regularly, where they interacted with the robot more than four times during the test week. Many of the other residents only interacted with the robot once.

We collected 14 consent forms from the personnel. We did not meet six of the staff members, but three regular caregivers were interested in the project and wanted to help. In addition to them, two volunteers, who did gymnastics with the residents one morning, wanted to interact with the robot as well and signed consent forms on the spot.

6.1.2 The Robot

The Robotic Service Assistant was developed by Fraunhofer IPA.[44] The drink-serving concept was investigated as part of the WiMi-Care project, using the Care-O-bot 3, which is a general service robot.[45] The robot has an omnidirectional mobile base, drink storage, a serving mechanism, and a touchscreen to facilitate interactions. On the tablet, the user can select a drink, or if the robot is not offering anything, the screen displays its eyes. The eyes are cartoon-like with black pupils/irises and turquoise sclera, and they are able to track movement autonomously.

Fig. 9: The drink serving robot

Figure 9 shows a person selecting a beverage on the touchscreen mounted on the robot. The robot provides the beverage in a cup that the participant then has to pick

44   Baumgarten/Jacobs/Graf, The robotic service assistant-relieving the nursing staff of work-load.

45   Theo Jacobs/Birgit Graf, Practical evaluation of service robots for support and routine tasks in an elderly care facility, in: *2012 IEEE Workshop on Advanced Robotics and its Social Impacts (ARSO)*, Munich 2020, 46–49.

up. The robot weighs around 150 kg and is approximately 1.5 m tall. The robot combines information from three lidar sensors and one RGBD camera to ensure safety during navigation. The RGBD camera also provides data for people detection.

6.1.2.1 Robot Dialogue

Since speech recognition is currently very hard to achieve, especially in interactions with older adults (unless the system is specifically trained on older adults' speech),[46] it was deemed too risky to rely on speech recognition in this trial.

Therefore, we created s cripted dialogues with the aim of ensuring that the older people did not need to say anything and only had to touch the screen to prompt the robot to serve the desired beverage.

The dialogue wizard manually played each of the robot's utterances. We created a set of functionally equivalent German utterances so that overhearers would not hear the same dialogues repeatedly. These included (translated) greetings, utterances to offer beverages, persuasive utterances (e.g., "Most women take apple juice, so you should do that too."), humorous utterances (involving various German wordplays), requests to touch the screen to order (e.g., "Please touch the picture of the desired beverage on the screen."), processing signals while the robot is serving the drink (e.g., "Please wait."), success utterances when the beverage is ready (e.g., "Please take your drink."), and closing salutations.

We also created a few utterances, as follows, that would be useful in case users pressed the wrong button:

- "If you accidentally touched the wrong picture, please let my assistant know. No worries."
- "Hoppela, did you really want water? One of my assistants will help you."

We anticipated that these utterances would be useful to have in the repertoire because the older persons were unfamiliar with this kind of technology. The utterances indeed proved to be helpful.

6.2   Findings

This study allowed us to identify numerous issues with robots in institutional interactions that the robot designers had not anticipated, for example, problems regarding the robot's appearance in the context of relating to this particular audience, the robot's voice, the touchscreen, and the beverage storage capacity. Furthermore,

---

46   Cf. Frank Rudzicz et al., Speech interaction with personal assistive robots supporting aging-at-home for individuals with Alzheimer's disease, in: *ACM Transactions on Accessible Computing* 7 (2/2015), 1–22.

our observations and in-depth interviews have revealed recommendations related to the interactions and, more generally, the design of service robots.

6.2.1 Serving Beverages

We conducted trials in which the robot offered beverages to residents sitting in one of the common areas as often as possible, as well as whenever the researchers and the caregivers thought it would fit the circumstances. The interactions generally had the following structure:

1) The navigation wizard decided which resident to approach, based on which resident was seated the closest to the robot's starting position. The robot then approached every person at the table, in turn, offering them something to drink, before continuing to the next table. When the robot arrived at a position next to a person, the dialogue wizard started a conversation with a greeting.

2) Next, the wizard played an utterance containing an offer. If the person did not hear the robot, the interaction could not continue. At this point, a researcher (first author) entered the interaction and repeated what the robot said. We had several interactions where an elder did not hear the robot.

3) In some cases, the wizard played a persuasive utterance, a wordplay, or a joke. (We did not want the dialogues to be repetitive or boring, so, as previously mentioned, we added these utterances.) The persuasive utterances were intended to persuade someone who declined an offer to have a drink after all. The wordplays and jokes were meant to entertain groups of people.

4) The robot then asked the participant to touch the picture of the desired beverage on the screen to initiate an order. We observed many difficulties regarding the use of the touchscreen. Not all the elderly residents could reach the screen because it was too high or too far away. In other cases, an elder could physically reach the screen but did not know how to use a touchscreen. Whenever the dialogue wizard saw that a participant had touched the screen (even though they did not successfully generate an order), she played an utterance confirming what she thought the order was intended to be.

5) Next, the robot prepared the beverage, which took some time. We observed long waiting times where no one knew whether the robot was doing anything at all.

6) When the beverage was ready for retrieval, an utterance was played to make the participant aware that they could take their beverage. Most participants were able to pick up their beverage independently.

7) Finally, the robot uttered a closing salutation and moved on to the next person.

6.2.2 Drink-Serving Robots in Institutions

In this section, we present the practical implications of our field trial, followed by the methodological implications.

At the facility, the residents were happy to be offered drinks they did not normally get, and thus, many residents took something from the robot and drank more than usual. In Study 1, we learned that the older adults in the Danish facility drink many different beverages in different containers. In comparison, the residents of the German facility drink water, and sparkling water and a yellow soft drink were always on the tables. Few residents used a personalized container, and those who needed one were infrequently present in the common areas where we conducted our field trial. From that perspective, the robot was indeed "adding value".

Regarding the robot's design, both the residents and the caregivers mentioned that the height of the robot (150 cm) and its weight (around 150 kg) are intimidating, especially when people are sitting down (90% of the residents were sitting during the interactions). Furthermore, the robot made a continuous sound that clashed with some residents' hearing aid devices. Notably, we found that one resident did not want to enter the common area when the robot was there because of such troubles with her hearing aid.

Another challenge that emerged was serving all residents promptly. To do this, the robot would need a much larger beverage storage capacity, as having to interrupt a serving session to be refilled was impractical and also resulted in unequal service provision to the residents. Some residents became frustrated when the robot arrived at their seat but could not provide them with the orange juice they wanted, for example. Moreover, the caregivers made jokes about the robot while we were refilling it, for instance, by telling the residents that they (the caregivers) would have been much faster at serving drinks.

These issues emerged especially in situations in which the robot was used at larger gatherings, which occurred due to the discovery of more suitable times to employ the robot than management initially suggested. That is, we found that the institutional workflow was quite different than anticipated. Before we arrived, we had asked which areas might be suited for drink serving and at which times drink serving would make sense. Management's suggestion was to mobilize the robot in the common areas where residents tend to sit between meals, but our findings suggested that other times may be more suitable, particularly when residents meet for their weekly exercise or weighing sessions. During these sessions, however, we faced a challenge concerning the robot's beverage storage capacity as the robot could not serve all residents promptly during the larger activities since it needed to be refilled (and reprogrammed).

Furthermore, as in Study 1, where we observed that the residents were very unfamiliar with smartphones, touchscreens, and tablets and needed assistance to use them, using the robot's touchscreen proved to be problematic here as well, even though the older adults in this care facility were younger on average and more agile than those in the Danish facility.

Next, we found that the language the researchers speak is quite important in these kinds of studies. If some of the researchers speak a foreign language (in our case, English), the mystery surrounding the robot becomes even greater, making it even harder for older persons to follow the proceedings. Elderly participants find it comforting to talk with the researchers and have them explain what they do. Therefore, it was very helpful that at least one person in our group (the first author) could talk to the older persons directly.

Another general finding concerns the long-term effects of field trials in elderly care facilities. Our observations show that the residents found the interruption of their daily lives tiresome, necessitating thorough precautions to ensure that interactions with robots in institutional settings are as seamless as possible. Ultimately, the residents were happy to see us, but they were not so happy to see the robot. In the interviews, they expressed that they had grown tired of the robot and that they found themselves easily irritated with it after a couple of days. Furthermore, when we talked to the residents and the caregivers, they stated that they could not see the necessity of this robotic technology because caregivers would be much faster at serving the beverages.

6.3   Discussion

The field trial yielded insights into real-world testing, and furthermore, it provided us with information on the usability of this particular drink-serving robot. We will now discuss several observations and suggest recommendations for future real-world testing.

We identified the long waiting time while the beverage was being prepared as problematic. One resident jokingly said that he would be close to dying of thirst before he received a drink from the robot,[47] and we also had several instances in which even we as researchers could not tell whether the robot was actively working. A simple solution would be improved feedback, such as a faster reaction time onscreen and a verbal utterance that is only played when the order has been accepted, which would require a connection between the utterances and the robot technology. Such a connection would be extremely useful in several other situations as well.[48]

Residents also experienced difficulty hearing the robot because its voice was not sufficiently loud. Whenever a resident failed to hear the robot, the interaction could not continue, and a researcher (first author) had to enter the interaction to repeat the robot's greeting or offer. After the robot had already spent two days at the facility, a resident who was hearing one of its utterances for the first time reacted with great

---

47   "Ich verdurste ja bevor ich etwas zu trinken bekomme".

48   E.g. Jakob Nielsen, Ten usability heuristics, URL: https://pdfs.semanticscholar.org/5f03/b25 1093aee730ab9772db2e1a8a7eb8522cb.pdf [last accessed: August 15, 2023].

surprise upon learning that the robot could speak. Moreover, even when residents heard the robot speaking, which was rare, the elders answered the robot at such a low volume that the dialogue wizard could not hear them, which meant that the wizard did not know when to play the next utterance or what it should be. The researcher had to repeat what the resident had said so that the interaction could continue. Hence, the researcher's participation was needed in all interactions, which rendered them unnatural and un-"real" as human-robot interactions.

As anticipated based on the previous study, we encountered problems with the use of the touchscreen. As described, when a resident decided which beverage they wanted, the robot asked them to touch the screen to place the order. Seated residents had difficulty reaching the tablet, and for those who could touch it, the screen was either too sensitive or not sensitive enough for the older persons to use.

During the trial, we were able to solve both problems, that is, with using the tablet and hearing the robot, because we had a researcher sitting next to the robot, who could press buttons on the touchscreen and repeat what the robot had said. In this capacity, the researcher acted as the robot's "voice." Although the researcher had intended to function as an observer, she, by necessity, became part of the interactions she was observing because her participation was the only way the robot could be used in those situations. Thus, assuming the role of facilitator in the human-robot interactions was the best solution from a practical perspective. Nonetheless, this changed the researcher's role as evidenced by one episode in which a resident ordered an orange juice from the researcher as soon as he saw her, which cast doubt on the usefulness of a drink-serving robot such a situation.

There were additional problems related to the wizards controlling the robot. The navigation wizard mobilized the robot with a joystick, and for security reasons, the wizard was positioned only 2 m behind the robot, which made it obvious that the robot was being controlled manually. Both residents and caregivers commented on this negatively or made fun of it. A solution could be to put more distance between the wizard and the robot and perhaps even hide the joystick from the participants' view.

Concerning the extent to which the field study reflected the institution's natural workflow, the fact that three to four researchers were present at all times attracted divergent comments from residents and caregivers. One resident opined that the technology was unnecessary since its use required so many people. Another resident responded to that by saying that she did not believe that we were doing anything related to the technology; she thought that we were merely observing and that the robot was largely autonomous. This exchange shows the vast differences in the residents' perceptions of both the field trial and the robot's capabilities. Afterwards, when we debriefed the management and the residents on our research, the caregivers were not surprised because the truth of the situation had been very obvious to them given that two researchers had been following the robot closely. However, researchers should still debrief the relevant parties after testing to ensure that people do not have incorrect expectations of robotic solutions.[49]

7. General Findings Gleaned From Entering the Field

We would like to share some ethical considerations and recommended preparations gleaned from our field trials that may be helpful to other researchers. Experiment participants generally need to sign consent forms. Collecting consent forms is very time-consuming and will be especially hard if the researchers do not live close to the given facility. We solved this problem by talking to the facility's management and arranging for the staff to collect the consent forms. However, in such cases, it might be difficult for facility managers and staff to explain research aspects about which they have little knowledge. For instance, our consent forms for Study 3 asked whether the participants would be willing to give short interviews. Many responded "No"; however, when the time for the interviews came, we found that several residents wanted to comment on the interactions even though they (or their relatives) had indicated otherwise on the consent forms. This discrepancy may have been because their understanding of an interview differed from ours. Additionally, relatives completing the form may have wanted to be cautious to prevent their parents, siblings, etc., from being endangered, which could have led to rather conservative choices on the consent form.

Many caregivers at both facilities were also reluctant to consent because they did not know what to expect. However, one caregiver at the German elderly care facility decided to participate after we had arrived because she realized that the trial was not frightening at all.

We also faced numerous comprehension problems regarding the need for consent forms. The caregivers and elderly persons were unfamiliar with basic research processes; for example, they did not understand the meaning of the phrase "pictures published in academic articles."[50] Such comprehension issues could cause researchers to lose participants if concepts are not sufficiently explained. To avoid this, researchers are advised to dedicate extra preparation time to personally collecting consent forms, so that they have the opportunity to explain what parties at participating facilities should expect.

Regarding recordings taken in care facilities, a general difficulty is recording only those persons who have consented. In the German elderly care facility, we ob-

---

49   Cf. Bertram F. Malle, Trust and the Discrepancy between Expectations and Actual Capabilities of Social Robots, in: Dan Zhang/Bin Wei (eds.), *Human-robot interaction: Control, analysis, and design*, New York 2020, 1–23.

50   This was one of the items participants needed to tick off on the consent forms.

served several instances where people from other floors who had not consented were visiting the area. Hence, researchers need to be aware of unforeseen participants and act spontaneously. During the studies conducted in the Danish elderly care facility, we had very few participants per unit, which made it hard for us to record any video data at all.

Moreover, it is important to remember that the facility is the residents' home, and researchers cannot make drastic changes to furniture or daily routines. Field trials are already quite intrusive because of the many unfamiliar people plus the robot that invade the residents' living spaces.

Another issue concerns the target population's specific characteristics. Some elderly residents may suffer from severe dementia, whereas others may be cognitively fit and "only" suffering from visual or hearing impairments. For instance, a Study 3 participant who wanted to interact with the robot could hear the robot but not see the screen. Thus, the design of robots for use with elders should feature good visual and hearing capabilities, which we had not considered prior to the testing. In Study 2, we only had one healthy participant who was able to interact with the robot. We noted much variance in this regard across the different care facilities. Some older persons may be fit for talking with a researcher but not fit to interact with a robot. It would be helpful to be aware of this distinction when conducting field trials, especially in elderly care facilities.

8. Conclusion

We presented three studies that were conducted in real-world environments, that is, in elderly care facilities. In Study 1, we observed human-robot interactions for a duration of 24 hours, which provided us with insights concerning users' real needs. In Study 2, we tested the navigation guidance task on one participant and identified several areas in which the interaction between the user and the robot needs to be improved. In Study 3, we visited another elderly care facility where a designated drink-serving robot was expected to be useful. However, the field trials highlighted several issues to be aware of when researchers conduct studies in real-world environments. Despite our efforts to prepare the parties at the facilities and ourselves for the field trials, we were unable to achieve true real-world testing because of the many factors that influence both how such studies can be conducted and the success of robot deployment in elderly care facilities. This paper has shed light on these factors and has offered some recommendations for other researchers who want to conduct real-world testing with real users.

Bibliography

Andriella, Antonio/Torras, Carme/Alenyà, Guillem, Short-term human–robot inter-action adaptability in real-world environments, in: *International Journal of Social Robotics* 12 (2020), 639–57.

Baumgarten, Simon/Jacobs, Theo/Graf, Birgit, The robotic service assistant-relieving the nursing staff of workload, in: *ISR 2018; 50th International Symposium on Robotics*, Munich 2018, 1–4.

Blomberg, Jeanette, Burrell, Mark, & Guest, G. (2003), An ethnographic approach to design, in: The Human-Computer Interaction Handbook. L. L: Erlbaum *Associates Inc., Hillsdale, NJ, USA.*

Bodenhagen, Leon/Fischer, Kerstin/Winther, Trine S./Langedijk, Rosalyn M./Skjøth, Mette M., Robot use cases for real needs: A large-scale ethnographic case study, in: *Journal of Behavioral Robotics* 10 (1/2019a), 193–206.

Bodenhagen, Leon/Suvei, Stefan-Daniel/Juel, William K./Brander, Erik/Krüger, Norbert, Robot technology for future welfare: meeting upcoming societal challenges–an outlook with offset in the development in Scandinavia, in: *Health and Technology* 9 (3/2019b), 197–218.

Bähler, Caroline/Huber, Carola A./Brüngger, Beat/Reich, Oliver, Multimorbidity, health care utilization and costs in an elderly community-dwelling population: a claims data based observational study, in: *BMC Health Services Research* 15 (1/2015), 1–12.

Chang, Wang-Ling/Šabanović, Selma/Huber, Lesa, Situated analysis of interactions between cognitively impaired older adults and the therapeutic robot PARO, in: *International Conference on Social Robotics* 2013, 371–380.

Chun, Bohkyung/Knight, Heather, The Robot Makers: An Ethnography of Anthropomorphism at a Robotics Company, in: *ACM Transactions on Human-Robot Interaction* 9 (3/2020), 1–36.

Fernaeus, Ylva/Håkansson, Maria/Jacobsson, Mattias/Ljungblad, Sara, How do you play with a robotic toy animal? A long-term study of Pleo, in: *IDC'10: Proceedings of the 9th international Conference on interaction Design and Children* 2010, 39–48.

Fink, Julia/Bauwens, Valérie/Kaplan, Frédéric, Kaplan/Dillenbourg, Pierre, Living with a vacuum cleaning robot, in: *International Journal of Social Robotics* 5 (3/2013), 389–408.

Fischer, Kerstin/Weigelin, Hanna M./Bodenhagen, Leon, Increasing trust in human–robot medical interactions: effects of transparency and adaptability, in: *Journal of Behavioral Robotics* 9 (1/2018), 95–109.

Fischer, Kerstin/Seibt, Johanna/Rodogno, Raffaele/Rasmussen, Majken/Weiss, Astrid/Juel, William K./Bodenhagen, Leon/Krüger, Norbert, Integrative Social Robotics Hands-On, in: *Interaction Studies* 21 (1/2020), 145–191.

Forlizzi, Jodi, How robotic products become social products: an ethnographic study of cleaning in the home, in: *HRI'07: Proceedings of the ACM/IEEE International Conference on Human-Robot Interaction* 2007, 129–136.

Hammersley, Martyn/Atkinson, Paul, *Ethnography: Principles in practice*, New York 2007.

Hasse, Cathrine/Trentemøller, Stine/Sorenson, Jessica, Special issue on ethnography in human-robot interaction research, in: *Journal of Behavioral Robotics* 10 (1/2019), 180–181.

Hebesberger, Denise/Körtner, Tobias/Pripfl, Jürgen/Gisinger, Christoph/Hanheide, Marc, What do staff in eldercare want a robot for? An assessment of potential tasks and user requirements for a long-term deployment, in: *IROS Workshop on "Bridging user needs to deployed applications of service robots*, Hamburg 2015.

Hoffman, Guy/Zhao, Xuan, A primer for conducting experiments in human–robot interaction, in: *ACM Transactions on Human-Robot Interaction* 10 (1/2020), 1–31.

Jacobs, Theo/Graf, Birgit, Practical evaluation of service robots for support and routine tasks in an elderly care facility, in: *2012 IEEE Workshop on Advanced Robotics and its Social Impacts (ARSO)*, Munich 2020, 46–49.

Juel, William K./Haarslev, Frederik/Ramirez, Eduardo R./Marchetti, Emanuela/Fischer, Kerstin/Shaikh, Danish/Manoonpong, Poramate/Hauch, Christian/Bodenhagen, Leon/Krüger, Norbert, Smooth robot: Design for a novel modular welfare robot, in: *Journal of Intelligent & Robotic Systems* 98 (1/2020), 19–37.

Jung, Malte/Hinds, Pamela, Robots in the wild: A time for more robust theories of human-robot interaction, in: *ACM Trans. Hum.-Robot Interact.* 7 (1/2018), Article 2.

Krüger, Norbert/Fischer, Kerstin/Manoonpong, Poramate/Palinko, Oskar/Bodenhagen, Leon/Baumann, Timo/Kjærum, Jens/Rano, Ignacio/Naik, Lakshadeep/Juel, William K./Haarslev, Frederik/Ignasov, Jevgeni/Marchetti, Emanuela/Langedijk, Rosalyn M./Kollakidou, Avgi/Jeppesen, Kasper C./Heidtmann, Conny/Dalgaard, Lars, The SMOOTH-robot: a modular, interactive service robot, in: *Frontiers in Robotics and AI* 8 (2021).

Kumar, Vijay/Whitney, Patrick, Faster, cheaper, deeper user research, in: *Design Management Journal (Former Series)* 14 (2/2003), 50–57.

Langedijk, Rosalyn M./Odabasi, Cagatay/Fischer, Kerstin/Graf, Birgit, Studying Drink-Serving Service Robots in the Real World, in: *2020 29th IEEE International Conference on Robot and Human Interactive Communication (RO-MAN)*, Naples 2020, 788–793.

Lee, Hee R./Cheong, EunJeong/Lim, Chaeyun/Fischer, Kerstin, Configuring Humans: What Roles Humans Play in HRI Research, in: *Proceedings of the IEEE Human-Robot Interaction Conference*, Sapporo 2022.

Mack, Natasha/Woodsong, Cynthia/MacQueen, Kathleen/Guest, Greg/Namey, Emily, *Qualitative research methods: A data collector's field guide*, North Carolina 2005.

Malle, Bertram F./Fischer, Kerstin/Young, James E./Moon, Ajung/Collins, Emily, Trust and the Discrepancy between Expectations and Actual Capabilities of Social Robots, in: Dan Zhang/Bin Wei (eds.), *Human-robot interaction: Control, analysis, and design*, New York 2020, 1–23.

Mutlu, Bilge/Forlizzi, Jodi, Robots in organizations: the role of workflow, social, and environmental factors in human-robot interaction, in: *HRI '08: Proceedings of the 3rd ACM/IEEE International Conference on Human-Robot Interaction* 2 (2020), 287–294.

Nielsen, Jakob, Ten usability heuristics, URL: https://pdfs.semanticscholar.org/5f0 3/b251093aee730ab9772db2e1a8a7eb8522cb.pdf [last accessed: August 15, 2023].

Reeves, Scott/Kuper, Ayelet/Hodges, Brian D., Qualitative research methodologies: ethnography. BMJ 2008, 337.

Rogers, Y./Sharp, H./Preece, J., Interaction design: beyond human-computer interaction (4th edition). New Jersey 2015.

Rudzicz, Frank/Wang, Rosalie/Begum, Momotaz/Mihailidis, Alex, Speech interaction with personal assistive robots supporting aging-at-home for individuals with Alzheimer's disease, in: *ACM Transactions on Accessible Computing* 7 (2/2015), 1–22.

Sabanovic, Selma/Michalowski, Marek P./Simmons, Reid, Robots in the wild: Observing human-robot social interaction outside the lab, in: *9th IEEE International Workshop on Advanced Motion Control*, Istanbul 2006, 596–601.

Sabelli, Alessandra M./Kanda, Takayuki/Hagita, Norihiro, A conversational robot in an elderly care center: an ethnographic study, in: *6th ACM/IEEE international conference on human-robot interaction*, 2011, 37–44.

Salvini, Pericle/Laschi, Cecilia/Dario, Paolo, Design for acceptability: improving robots' coexistence in human society, in: *International journal of social robotics* 2 (4/2010), 451–460.

Seibt, Johanna/Damholdt, Malene F./Vestergaard, Christina, Integrative social robotics, value-driven design, and transdisciplinarity, in: *Interaction Studies* 21 (1/2020), 111–144.

Spradley, James P., *Participant observation*, New York 1980.

Sung, JaYoung/Christensen, Henrik I./Grinter, Rebecca E., Robots in the wild: understanding long-term use, in: *Proceedings of the 4th ACM/IEEE international conference on Human robot interaction*, 2009, 45–52.

Veling, Louise/McGinn, Conor, Qualitative Research in HRI: A Review and Taxonomy, in: *International Journal of Social Robotics* 13 (2021), 1–21.

Wasson, Christina, Ethnography in the field of design, in: *Human organization* 59 (4/2000), 377–388.

Part II: Embodied Agents in (Inter-)Action

Mixed Methods for Mixed Realities: The Analysis of Multimodal Interactions With Embodied Conversational Agents

Jonathan Harth

Abstract *Multimodal interactions with anthropomorphic virtual agents are increasingly becoming the subject of current research on human–agent interaction, but existing research paradigms often focus only on the user's perception of interactions and do not position the emergent interaction processes themselves as the research object. The methodological approach presented here addresses this problem and focuses on both the relationship level as well as the content level in human–agent interaction. The combination of mixed reality representations and mixed methods allows, among other things, for the identification of possible discrepancies between the user's individual experiences and their physically expressed behavior during these interactions.*

1. Introduction[1]

Interaction with embodied conversational agents is increasingly becoming a common part of many people's everyday lives as technology advances.[2] These virtual assistants aim to act as a natural human–computer interaction (HCI) interface. The use of these interfaces is primarily seen in the business or administrative context (e.g., customer service, expert systems, and more), and they promise natural, multimodal conversations in real time, using both verbal and nonverbal expressions.[3] See,

---

1   This research was made possible by funding from the European Regional Development Fund (ERDF).
2   Cf. James N. Weinstein, Artificial Intelligence: Have You Met Your New Friends; Siri, Cortana, Alexa, Dot, Spot, and Puck, in: *Spine* 44 (1/2019), 1–4.
3   Farina Freigang/Sören Klett/Stefan Kopp, Pragmatic Multimodality: Effects of Nonverbal Cues of Focus and Certainty in a Virtual Human, in: Jonas Beskow et al. (eds.), *Intelligent Virtual Agents*, 17th International Conference, IVA 2017, Stockholm, Sweden, August 27–30, 2017, Proceedings, 142–155.

for example, the company Digital Human,[4] which aims to make interacting with conversational agents more natural and realistic. Another major company, SoulMachines, defines itself as "the world leader in humanizing AI to create astonishing Digital People" that are able to "engage your customers in a powerful new way."[5] The goal of such endeavors is the development of intelligent products that enable organizations to rapidly scale customer brand experiences.[6]

In addition to voice only, virtual humans are enabled for nonverbal interactions. These so-called *embodied conversational agents* (ECAs) can use facial expressions, gestures, and physical attention to convey information, in addition to engaging purely verbally in exchanging information.[7] Current ECAs, such as Virtual Mike,[8] Mica,[9] and Digital Douglas,[10] illustrate the rapid development of such systems, which combine natural language processing and embodiment.

The basic prerequisite, however, is a high-quality interaction interface that gives the user the feeling of an intelligent and complex dialogue partner.[11] Research on human–robot interaction (HRI) encompasses these conditions. In HRI, the focus has shifted from speech as the most obvious communication mode to nonverbal cues, but this shift has created new challenges in the technical development of systems as well as in the analysis of the resulting interactions' complexity.[12] Challenges also exist for human–agent interaction (HAI): The processing and generation of nonverbal communications lead to new implications for the design, development, and evaluation of interaction processes between humans and virtual agents.[13] Although we frequently read claims of "super-human performance" in speech recognition, image processing, and so forth, "no system," as Kopp and Krämer recently stated, "is

---

4   Digital Humans Inc., URL: https://digitalhumans.com/ [last accessed: August 15, 2023].

5   SoulMachines, URL: https://www.soulmachines.com/ [last accessed: August 15, 2023].

6   Fred Miao et al., EXPRESS: An Emerging Theory of Avatar Marketing, in: *Journal of Marketing* (4/2021).

7   Justine Cassel et al., *Embodied Conversational Agents*, Cambridge; Massachusetts; London 2000.

8   Mike Seymour/Chris Evans/Kim Libreri, Meet Mike: epic avatars, in: *ACM SIGGRAPH 2017 VR Village (SIGGRAPH '17)*. Association for Computing Machinery, New York 2017, Article 12, 1–2.

9   Magic Leap Inc, *Magic Leap's Mica at GDC*, URL: https://www.youtube.com/watch?v=-PzeWxt OGzQ [last accessed: August 15, 2023].

10   Digital Domain, *Introducing Douglas—Autonomous Digital Human*, URL: https://www.youtub e.com/watch?v=RKiGfGQxqaQ [last accessed: August 15, 2023].

11   Li Gong, How social is social responses to computers? The function of the degree of anthropomorphism in computer representations, in: *Computers in Human Behavior* 24 (4/2008), 1494–1509.

12   Christoph Bartneck et al., *Human-Robot Interaction. An Introduction*, Cambridge 2019.

13   Jan A. Deriu et al., Survey on Evaluation Methods for Dialogue Systems, in: *arXiv*:1905.04071v2, 2020.

able to lead a half-decent coherent conversation with a human."[14] However, the recent emergence of powerful large language models, such as GPT-3, have, for the first time, realized a much higher level of coherence and plausibility in communication.[15]

The use of virtual reality (VR) in research on interactions with ECAs should help in at least three regards. First, users immersed in VR can fully commit to the situation. Compared to studies dedicated to exchange with agents or chatbots on computer screens, VR succeeds in shedding such a situation's "two-worldliness."[16] Although typical studies are structured in such a way that users sit in front of a computer monitor to interact with an agent "in" the computer, with VR, the subjects visit the agent in "its" habitat. This characteristic leads to the second notable merit as VR enables virtual interactions with virtual humans face-to-face, which is the "gold standard" for research on interactions. As Bavelas et al. have pointed out, face-to-face conversations incorporate several features that distinguish the mode from other forms of communication, including (a) unrestricted verbal communication, (b) full nonverbal communication on all channels, and (c) continuous coordination among the conversational partners. However, a crucial question here is whether this virtual face-to-face situation is "symmetrical" in the sense that both participants have the same perceptual and expressive abilities. We will return to this. The third merit is that recent VR technology can wholly utilize full body tracking.[17] Through advanced controllers and body tracking, even sublime user body movements can be captured—and potentially made accessible to the virtual agent.[18]

However, for social sciences that aim at answering the questions of conversational agents' usability, plausibility, and acceptance, these kinds of interactions represent a methodological challenge first and foremost.[19] VR studies encounter the problem that interactions are not easily observable "from outside" the VR headset

14   Stefan Kopp/Nicole Krämer, Revisiting Human-Agent Communication: The Importance of Joint Co-construction and Understanding Mental States, in: *Frontiers of Psychology* 12 (2021), 580955.

15   Jonathan Harth/Martin Feißt, Neue soziale Kontingenzmaschinen. Überlegungen zu künstlicher sozialer Intelligenz am Beispiel der Interaktion mit GPT-3. In: Schnell, Martin (eds.): *Begegnungen mit künstlicher Intelligenz. Intersubjektivität, Technik, Lebenswelt*, Weilerswist 2022.

16   Antonia Krummheuer, *Interaktion mit virtuellen Agenten? Zur Aneignung eines ungewohnten Artefakts*, Stuttgart, 2010.

17   Janet B. Bavelas et al., Using face-to-face dialogue as a standard for other communication systems, in: *Canadian Journal of Communication* 22 (1/1997), 5.

18   Mary Ellen Foster, Face-to-Face Conversation: Why Embodiment Matters for Conversational User Interfaces, in: *Proceedings of the 1st International Conference on Conversational User Interfaces (CUI '19)*. Association for Computing Machinery, New York 2019, Article 13, 1–3.

19   Siska Fitrianie et al., What are We Measuring Anyway? A Literature Survey of Questionnaires Used in Studies Reported in the Intelligent Virtual Agent Conferences, in: *Proceedings of the 19th ACM International Conference on Intelligent Virtual Agents* (2019), 159–161.

and are mostly recorded solely from the user's perspective. However, the possibilities of mixed reality representation offer a solution to this problem. This paper will present the approach and apply it to the analysis of multimodal interactions with ECAs.

The following paper presents a possible solution to the increasingly pressing future need to study interactions between human users and virtual agents in VR. The outline of this approach is illustrated by examples derived from an ongoing research and development project called Ai.vatar—The Virtual Intelligent Assistant (EFRE, IT-2-2-030c). First, the communicative possibilities and limitations of currently available ECAs are presented and discussed, demonstrating that these ECAs tend to communicate exclusively on the content level of communication while neglecting the relationship level. Next, selected methodological challenges that accompany the analysis of the relational level in interactions with ECAs will be presented. The following chapter is dedicated to the methodological elaboration of individual steps comprising mixed reality methods. The materials and procedures will be presented, as well as the specific analysis of multimodal interactions. This chapter ends with an exemplary presentation of a transcription and interpretation. In the concluding section, further implications and possibilities for extending the presented method will be discussed.

2. ECAs' Communicative Capabilities

For a long time, the development of realistic, lifelike embodied agents has been mired in the so-called "uncanny valley."[20] Current agents use facial micro-expressions, gestures, and plausible dialogue skills that aim at simulating humanness.[21] Consequently, users increasingly perceive conversational agents as humanlike.[22] Developments with regard to visual fidelity, text-to-speech and speech-to-text algorithms, as well as dialogue skills have led to an increased focus on nonverbal communication as an option for rich HAI. The reason for this is obvious: Not only do users prefer embodied, realistic, and humanlike visualizations of agents to voice

---

20   Masahiro Mori/Karl F. MacDorman/Norri Kageki, The Uncanny Valley, in: *IEEE Robotics & Automation Magazine* 19 (2012), 98–100.

21   David Burden/Maggi Savin-Baden, *Virtual Humans. Today and Tomorrow*, Boca Raton 2020.

22   See Philip R. Doyle et al., Mapping Perceptions of Humanness in Intelligent Personal Assistant Interaction, in: *Proceedings of the 21st International Conference on Human-Computer Interaction with Mobile Devices and Services*, 2019., See also Ryan Lowe et al., Towards an automatic turing test: Learning to evaluate dialogue responses; in: arXiv:1708.07149v2, 2018.

only,[23] but ECAs can also promote intuitive understanding while leading to greater connectedness.[24]

However, although conversational agents have already attained a high level of natural language processing,[25] humans are still far superior to virtual agents at processing multimodal information. In addition to using speech, humans use gestures, paraverbal and facial expressions, as well as more or less expressive body postures for communication.[26] In the domain of nonverbal expressions, current virtual agents are only able to express themselves on a very basic level and usually completely lack the competence to process those nonverbal messages emitted by human users.[27]

In addition to this more or less technical challenge, another problem lies within nonverbal communication in general. According to Watzlawick et al., nonverbal expressions are always ambiguous:

> There are tears of sorrow and tears of joy, the clenched fist may signal aggression or constraint, a smile may convey sympathy or contempt, reticence can be interpreted as tactfulness or indifference, and we wonder if perhaps all analogic messages have this curiously ambiguous quality.[28]

This thinking led Watzlawick et al. to distinguish between digital and analog modalities of communication: *Digital codes* refer to what a person says and *what* the words actually mean, whereas *analog codes* refer to *how* something is said. Consequently, a sender can convey two contradictory messages at once, which raises the question of how exactly virtual agents should decode users' nonverbal actions, even if they could technically process this information.

Transferred to today's HAI, this means that humans and agents currently communicate exclusively on the content level, and the relationship level remains mainly unused or even obscured. How an agent is to understand a voice message—whether as an instruction, an assurance, or neutral information, etc.—can be read only

---

23   Jens Reinhardt/Luca Hillen/Katrin Wolf, Embedding Conversational Agents into AR: Invisible or with a Realistic Human Body? in: *Proceedings of the Fourteenth International Conference on Tangible, Embedded, and Embodied Interaction*, ACM Press, New York 2020.

24   Hung-Hsuan Huang, Embodied conversational agents, in: K. L. Norman & J. Kirakowski (eds.), *The Wiley handbook of human computer interaction*, Blackwell 2018, 601–614.

25   Daniel Adiwardana et al., Towards a Human-like Open-Domain Chatbot, in: *arXiv:2001.09977*, 2020.

26   Paul Watzlawick/Janet H. Beavin/Doti D. Jackson, *Pragmatics of Human Communication. A Study of Interactional Patterns, Pathologies, and Paradoxes*, New York 1967.

27   Benjamin Weiss et al., Evaluating embodied conversational agents in multimodal interfaces, in: *Computational Cognitive Science* 1 (6/2015).

28   Watzlawick/Beavin/Jackson, *Pragmatics of Human Communication*.

poorly from the plain linguistics of a communication. The important contextual meaning of linguistic information can only be supported, with difficulty, using words. Social reality is precarious, as ethnomethodology has argued. Social order and structure ultimately exist only in the form of temporary arrangements that are susceptible to disruption and must always be renewed, changed, and "repaired."[29] Although specialized algorithms are already able to identify facial expressions in terms of probable emotional expressions, the situation is quite different when it comes to observing a user's hand movements, snorting, intonation, or body postures and discerning the possible meanings of those messages. Consequently, with today's technology, the human user is usually far superior to the virtual agent in terms of decoding messages. In case of misunderstandings, so far, only the user has the chance to repair the communication, and the kind of communicative repair to choose,[30] for example, repetition, elaboration, or even topic shifts, is up to the user.[31]

From this perspective, current HAI appears to be almost "relationship-less." Consequently, we currently face the problem of a large discrepancy between the ability to communicate digitally (i.e., verbally, content-related) and the ability to communicate analogously (i.e., nonverbally, relationship-related) in interactions with virtual agents. It is precisely this deficit that becomes evident in studies that encounter an "uncanny valley" in HAI.[32] Interaction with virtual agents "lacks" something—namely, the relationship level: "One gesture or facial expression tells us more about how another person thinks about us than a hundred words."[33] This effect leads to increased error-proneness in HAI because contextualization of the framing situation as well as the specific user–agent dyad are not processed.[34] Therefore, while ECAs are increasingly excelling in the area of processing spoken language, their deficit in the area of processing relational cues and nonverbal communication is becoming more and more apparent. The use of the VR medium can at least open up new ways for tackling this since full body tracking can provide a joint bodily co-presence.

---

29   Makoto Hayashi, Geoffrey Raymond, Jack Sidnell (eds.), *Conversational Repair and Human Understanding*, Cambridge 1992.

30   Hedda Meadan/James W. Halle, Communication Repair and Response Classes, in: *The Behavior Analyst Today* 5 (3/2004).

31   Mark Dingemanse et al., Universal Principles in the Repair of Communication Problems, in: *PLOS ONE* 2015, e0136100.

32   Markus Thaler/Stephan Schlögl/Aleksander Groth, Agent vs. Avatar: Comparing Embodied Conversational Agents Concerning Characteristics of the Uncanny Valley, in: *IEEE International Conference on Human-Machine Systems (ICHMS)*, 2020.

33   Watzlawick/Beavin/Jackson, *Pragmatics of Human Communication*, 64.

34   Weiss et al., *Evaluating embodied conversational agents in multimodal interfaces*.

3. Methodological Challenges for Studying the Relationship Level in HAI

Evidently, there is no generally shared consensus on methodological approaches, not only in the field of VR research but also in research on HAI.[35] Moreover, research must grapple with a multitude of open questions regarding methodological, technical, social, and other challenges that require a focused investigation.[36] Thus, although ECAs are becoming increasingly sophisticated and are introducing more complex multimodal information to interactions, existing methodological approaches often lack the necessary tools to deal with this. A recent meta-analysis of instruments used in HAI has shown that the question of the "relationship" between the user and the agent usually remains untouched.[37] It seems as if the social dimension of interaction is mostly overlooked in the predominantly psychologically motivated research on HAI. Usually, the agent's "impression" of the user is not considered, and the interaction is only analyzed in terms of jointly coordinated behavior. The methodological approach elucidated here comes into play in exactly this blind spot. In conjunction with Kopp and Krämer,[38] this paper suggests bringing about a sociological turn in the research on interactions with ECAs by studying co-produced joint behavior during these interactions.

For this, we can draw on a vast corpus of existing and well-established methodologies for analyzing interactions themselves. A large number of studies conducted in the last decades can be regarded as sociological research on more or less formalized interactions. Consequently, we have well-established ways to analyze different types of interactions: conversation analysis, ethnomethodology, and workplace studies. As Mondada recently stated, these kinds "of analysis [have] made it possible to identify the specific sequential formats that configure and constrain the opportunities to speak and to initiate actions—as they are shaped by the institutional context but also reflexively construct it—in a number of institutional interactions."[39]

Adapting such an approach to HAI would mean focusing on how human users and virtual agents ceaselessly draw upon interactional rules and practices when constructing more or less shared understandings of what is unfolding within the interaction. According to Heritage, a "reflexive dimension" in social action is central to

---

35   Fitrianie et al., *What are We Measuring Anyway?*

36   Dmitry Alexandrovsky et al., Evaluationg User Experiences in Mixed Reality, in: *arXiv* preprint arXiv:2101.06444, 2021.

37   Siska Fitrianie et al., The 19 Unifying Questionnaire Constructs of Artificial Social Agents: An IVA Community Analysis, in: *Proceedings of the 20th ACM International Conference on Intelligent Virtual Agents (IVA '20)*. Association for Computing Machinery, New York 2020, Article 21, 1–8.

38   Kopp/Krämer, *Revisiting Human-Agent Communication.*

39   Lorenza Mondada, Conversation Analysis and Institutional Interaction, in: *The Encyclopedia of Applied Linguistics* (2012).

this process: By "their actions[,] participants exhibit an analysis or an understanding of the event in which they are engaged, but by acting[,] they also make an interactional contribution that moves the event itself forward on the basis of that analysis."[40] For this reflexive dimension, it is necessary to identify appropriate cues, which help with reconstructing the interactional rules and practices. Here, we can draw from research on HAI because, even in the earliest social robot designs, nonverbal cues have been used to enrich interactions.[41] For example, even the early social robot Kismet was able to use different body postures to express affect and engage people in interaction.[42]

In all interactions, nonverbal cues are delivered in several modalities at once; however, for analysis, it might be worthwhile to consider each type of nonverbal cue separately. According to Bartneck et al., we can summarize these nonverbal cues as (a) gaze and eye movement, (b) gesture, (c) mimicry and imitation, (d) touch, (e) posture and movement, and (f) interaction rhythm and timing.[43] With this approach, we can reconstruct verbal and nonverbal interactional patterns and practices in HAI that might even be contradictory. For example, users may verbally express positive ideas about the agent whereas their nonverbal cues indicate disapproval or even disrespect towards the agent.

According to Bartneck et al., nonverbal cues can be used as indications for whether a user is enjoying interacting with a robot—or in our case, with a virtual agent. In accordance with Watzlawick et al., nonverbal communication has to be regarded as always rooted in a specific context, as it is the context that renders a given nonverbal signal appropriate (or not). It is exactly these contexts that guide the meaning of verbal and nonverbal information. For reconstructive methodologies such as conversation analysis or interaction analysis, which focus on interactional rules, patterns, and practices, a focus on nonverbal cues should help the analysis. That is why multimodal interaction analysis can be understood as a further development of conversation analysis against the backdrop of new technical possibilities. Whereas conversation analysis originally focused on spoken language alone, in

---

40   John Heritage, Conversation Analysis and Institutional Talk: Analyzing Distinctive Turn-Taking Systems, in: S. Cmejrková et al. (eds.), *Proceedings of the 6th International Congresss of IADA (International Association for Dialog Analysis)*, Tübingen 1998, 3–17.

41   Andreas Bischof, *Soziale Maschinen bauen. Epistemische Praktiken der Sozialrobotik*, Bielefeld 2017.

42   Cynthia Breazeal, Toward sociable robots, in: *Robotics and Autonomous Systems* 42 (3–4/2003), 167–175.

43   Bartneck et al., *Human-Robot Interaction*.

recent years, it has increasingly turned towards multimodality, which is why it can now also be used very effectively for the analysis of nonverbal communication.[44]

4. Mixed Reality Methods for the Analysis of Multimodal Interactions

Based on the theoretical and methodological background mentioned before, we would like to present the idea of a reconstructive approach that uses mixed reality methods for analyzing multimodal HAI. Overall, the methodological approach is characterized by a very high degree of openness. Consequently, the use of mixed reality methods should facilitate the collection of different types of data: We can still measure the user's assessment of an agent using questionnaires and interviews, but we are now able to observe the jointly co-produced interaction of the user and the agent. Further, in addition to verbal cues alone, we can evaluate nonverbal cues as well. From this, we hope to gain insights into *different modes of interactions* on both the content and the relationship levels, which may point towards general patterns of interaction with virtual humans.

4.1   Materials and Procedures

The ECA used for the outline of this methodological paper is currently being developed as part of the research project Ai.vatar. The agent can process spoken language as input and produce output using natural language processing via Google DialogFlow. This process is guided by an individually designed dialog management system that is successively extended by further interaction modalities. The goal is to integrate eye-tracking information as well as further generative speech generation (i.e., GPT-3) into the system while the project is still running. On the current state of the art of the prototype (see Fig. 1), users can communicate with the agent by speaking freely into the VR headset's microphone. The agent's appearance was created via photogrammetry. The agent's body is fully rigged and animated in Unreal Engine 4.27. All technical features are realized by the project partners IOX GmbH and HHVision GbR.

---

44   Henning Mayer/Florian Muhle/Indra Bock, Whiteboxing MAX. Zur äußeren und inneren Interaktionsarchitektur eines virtuellen Agenten, in: Eckhard Geitz/Christian Vater/Silke Zimmer (eds.), *Black Boxes—Versiegelungskontexte und Öffnungsversuche*, Berlin 2020, 295–322.

Fig. 1: Static rendering of the virtual agent "Florian"

The unique feature of the methodological approach presented here is that users are filmed on video camera during the entire interaction. The camera feed is live-matched with virtual images from Unreal Engine for a mixed reality representation. Mixed reality recording is enabled by implementing the LIV Mixed Reality Software Developer Kit (SDK) directly in the application.[45]

4.2   Multimodal Interaction Analysis

The methodological approach is rooted in the tradition of science and technology studies (more precisely, workplace studies), which investigate the interactive and situational processes involved in dealing with computer-based media and technical artifacts. The corresponding methods for this approach are based on concepts from ethnography, ethnomethodology, and conversation analysis. These concepts have proven successful in both focusing on the situational unfolding of mediated communication and identifying potential incongruities between the interpretive patterns of human users and the (pre-)determined structures of technical artifacts.[46] In the following paragraphs, we mainly follow the methodological concepts of conversation analysis, with the minor difference that, unlike classical workplace studies,

---

45   LIV, URL: https://liv.tv/ [last accessed: August 15, 2023].
46   Lucy Suchman, *Human-Machine Reconfigurations. Plans and Situated Actions* (2nd edition), Cambridge 2007.

we are not dealing with routine (work) processes when interacting with the virtual agent but rather with the adoption of an "unfamiliar artifact."[47]

At least since Suchman's landmark study,[48] the methods of ethnography and conversation analysis have become established as methods for researching human–machine interactions. With workplace studies, a research tradition has emerged that empirically investigates the situational and interactive use of technology, and conversation analysis has already proven its worth in the still young fields of HRI and HAI. See, for example, the execution of conversation analysis of HRI.[49]

An important tenet of conversation analysis is that it focuses on the *situatedness* of interactions. To perform conversation analysis, it is important to consider that every interaction takes place in a context.[50] Moreover, the relation between interaction and context must also be described as "reflexive."[51] Thus, conversation analysis analyzes audiovisual recordings of interactions in terms of their context-independent as well as context-dependent organizing principles or the rules and patterns by which its participants make social sense of the situation.[52] The goal of conversation analysis is to reconstruct the interactive patterns and organizing structures of interactions. Therefore, for conversation analysis, the focus is primarily on the *communicative events* between humans and machines and not, for example, on the question of subjective interpretations of the technical artifact. That kind of approach resembles modern concepts of communication theory also focused primarily on observing *communication* without relying on assessing users' expectations regarding agents' or robots' supposed properties.[53]

In particular, ethnomethodologically informed conversation analysis does not fall back on a predefined apparatus of methods but rather orients itself to Garfinkel's

47   Antonia Krummheuer, Conversation Analysis, Video Recordings, and Human-Computer Interchanges, in: Ulrike T. Kissmann (ed.), *Video Interaction Analysis. Methods and Methodology*, Frankfurt a. M. 2009, 59–83.

48   Lucy Suchman, *Plans and situated actions: The problem of human-machine communication*, Cambridge 1987.

49   See, e.g., Indra Bock/Henning Mayer, Humanoide Roboter und virtuelle Agenten als Kommunikationsteilnehmer? Konversationsanalytische Studien der Mensch-Maschine-Interaktion, in: Ahner, H., Metzger, M., & Nolte, M. (eds), *Von Menschen und Maschinen: Interdisziplinäre Perspektiven auf das Verhältnis von Gesellschaft und Technik in Vergangenheit, Gegenwart und Zukunft*, 2020. See Mayer/Muhle/Bock, *Whiteboxing MAX*.

50   Erving Goffman, *Encounters. Two Studies in the Sociology of Interaction*, Hamrondsworth 1972.

51   Harold Garfinkel, *Studies in Ethnomethodology*, Englewood Cliffs 1967.

52   Harvey Sacks/Emanuel A. Schegloff/Gail Jefferson, A Simplest Systematics for the Organization of Turn-Taking for Conversation, in: *Language* 50 (4/1974), 696–735.

53   Cf. Florian Muhle, Sozialität von und mit Robotern? Drei soziologische Antworten und eine kommunikationstheoretische Alternative, in: *Zeitschrift für Soziologie* 47 (3/2018), 147–163. See also Florian Muhle, Humanoide Roboter als 'technische Adressen'. Zur Rekonstruktion einer Mensch-Roboter-Begegnung im Museum, in: *Sozialer Sinn* 20 (1/2019), 85–128.

postulate of the "unique adequacy of methods."[54] This means that conversation analysis should always be done in a way that is appropriate to the subject's matter. Nevertheless, we can also give some general principles according to which conversation analysis proceeds. As a first step, audiovisual recordings of an interaction are made, which are then used as a basis for analysis. The recordings capture the situation in its temporal course and thus enable a precise analysis of the interaction processes. This is done by transcribing video and audio material. According to Sacks, the creation of the transcripts should be guided by the premise that "there is order at all points."[55] This is to express that the interaction sequence is not seen as random but rather as an expression of a more or less latent social order produced by the participants during the process itself. A typical challenge in producing transcripts of audiovisual data lies in the fact that a compromise must be found between practicable readability and detailed reproduction of the events.

Multimodal interaction analysis operates based on audiovisual interaction documents and focuses on conversational aspects such as turn-taking, interruptions, and modes of expression, as well as on extended modalities such as gaze behavior, gesticulation, facial expressions, and so on. Therefore, multimodal interaction analysis primarily uses video recordings to reconstruct the structures of interaction.[56] Of particular interest is the *practical* structuring of the interactive exchange between a human user and a virtual agent. Here, the primary focus lies on the organization of speaker changes, nonverbal behavior, and the handling of possible interferences. All patterns and structures are then analyzed with regard to their social (i.e., communicative) functions. For analysis in our planned studies, we identified five possible sequences that are of particular interest:

- The startup sequence of the interaction in VR
- The initial reaction to the agent
- The responses to the agent's questions and answers
- Possible addressing of the experimenters in the lab
- The closing of the conversation and goodbyes

For analysis of the relationship level in communicative interaction, it is important to examine the data to identify patterns that contain implicit or explicit indications

---

54   Harold Garfinkel, *Ethnomethodology's Program. Working out Durkheim's Aphorism*, Lanham 2002.

55   Harvey Sacks, Notes on Methodology, in: J.M. Atkinson/J. Heritage (eds.), *Structures of Social Actions. Studies in Conversation Analysis*, Cambridge 1984, 21–27.

56   Reinhold Schmitt, Positionspapier: Multimodale Interaktionsanalyse, in: Ulrich Dausend-schön-Gay/Elisabeth Gülich/Ulrich Krafft (eds.), *Ko-Konstruktionen in der Interaktion: Die gemeinsame Arbeit an Äußerungen und anderen sozialen Ereignissen*, Bielefeld 2015, 43–51.

regarding the relationship. Following Watzlawick et al., with just the integration of gestural expressions, such as gestures in opposition to verbal expressions, we can build a better understanding of the relationship level that is present in every inter- action. For example, it is possible to say that one feels positively about something while simultaneously making a face so that what is said takes on a completely differ- ent meaning. Even if attempts are made to avoid communication, exactly this would have to be communicated as some interactive behavior. Watzlawick et al. themselves have given an example of how the manner of communication can determine how a piece of information is understood: In their example, woman A points to woman B's necklace and asks, "Are those real pearls?" On this, the scholars wrote:

> [T]he content of her question is a request for information about an object. But at the same time she also gives—indeed, cannot *not* give—her definition of their relationship. How she asks (especially, in this case, the tone and stress of voice, facial expression, and context) would indicate comfortable friendliness, competi- tiveness, formal business relations, etc. B can accept, reject or redefine but cannot under any circumstances—even by silence—not respond to A's message.[57]

Therefore, it is especially important to focus on peculiar, surprising, or deviant pat- terns in interactions since these indications could be used to decode the ambiguity of analog communication and thus help to understand the jointly constructed rela- tionship between the interactive agents.

4.3   Transcription of Mixed Reality Data

Compared to merely auditory data, audiovisual data place special demands on tran- scription.[58] Participants' nonverbal actions especially come into focus. Here, it is necessary to notate both the verbal and nonverbal utterances and actions during the encounter, such as laughter and throat clearing, as well as any pauses, addressing of the experimenters in the room, terminations of words, synchronizations, mutual interruptions, or possible accentuations. According to Bartneck et al., we have to as- sume different types of nonverbal cues[59] that may occur in HRI: gaze and eye move- ment, gesture, mimicry and imitation, touch, posture and movement, interaction rhythm and timing, as well as other social mechanisms of face-to-face interactions

---

57   Paul Watzlawick/Janet Beavin, Some Formal Aspects of Communication, in: *The American Be- havioral Scientist* 10 (8/1967), 4–8.
58   Hubert Knoblauch/René Tuma, Videography: an interpretive approach to video-recorded micro-social interaction, in: Eric Margolis/Luc Pauwels (eds.), *The Sage Handbook of Visual Methods*, Thousand Oaks 2011, 414–430.
59   Bartneck et al., *Human-Robot Interaction*.

such as repairing failed sense-making.[60] With the exception of touch, we can apply all of these types to HAI as well. Accordingly, the first task in transcription is to provide information on these levels regarding the gestures, glances, body rhythms and timings, and so on.

Additionally, Lucy Suchman pointed out that it can be useful to combine the transcription with the respective perspectivity of the actors involved. Here, Suchman has recommended a tabular form for the transcription of human–machine interactions[61] that should be based on whether the actions of the user and the machine are transparent or accessible to the other (see Table 1).

Table 1: Suchman's Perspectivation of Interaction Processes

user	user	machine	machine
Situational interpretations of the human users that are *not accessible* to the machine.	Actions of the human users that *are accessible* to the machine.	Actions of the machine that *are accessible* to the human users.	Machine processes that are *not accessible* to the human users.

If we apply Suchman's analysis scheme to interaction with the agent from our research project, the gross asymmetry in the mutual perception becomes immediately clear. While the user can understand the agent's statements on both the informational and the communicative levels (i.e., the levels of *what* is said and *how* it is said can be related), the agent only has access to the informational level of communication. Due to the limitations of Dialogflow, the communicative exchange is only interpreted in the digital dimension. Thus, the agent is unable to interpret the user's way of speaking (as in sentiment analysis), and the current prototype cannot perceive the user's bodily gestures, postures, or facial expressions.

The situation appears to be quite different on the user's side since many of the agent's physical expressions are available to the user for interpretation: The virtual agent randomly produces bodily poses and hand movements (e.g., crossing the arms in front of the chest, raising one hand, looking at the hands). Additionally, the agent is able to (independently) align both its head and eyes with the user's position.

In comparison to Suchman's *perspectivation* of the actors (user and machine) involved in a situation, we are additionally interested in the *interaction process* between user and agent. Thus, we are interested in the processes of generating,

---

60   Jack Sidnell, *Conversation analysis. An introduction*, Chichester 2010.
61   Suchman, *Plans and situated actions*.

structuring, and evolving the interactive situations as shared between *both actors*. This ethnomethodological perspective is more akin to the third person perspective of an external observer who witnesses and subsequently attempts to interpret what is happening between the two. Therefore, we transcribe all activities of both actors in chronological order. The advantage of this approach is that any simultaneously occurring actions can be noted in an appropriate form and do not have to be sequentially fixed. This way, the events can be read *diachronically* from left to right and *synchronously* from top to bottom.

The participants' verbal utterances as well as all physical actions and nonverbal expressions that appear to be of relevance should be noted in the transcriptions. The verbal expressions are notated according to common conventions, and the nonverbal activities are placed in parentheses to identify them as an interpreter's initial observations. In addition to plain text transcription, in some cases, it will be necessary to include selected screenshots of the mixed reality video in the transcripts. In the following, we will present an example of such transcripts to illustrate the procedures. The audiovisual data used for this purpose were obtained from a (technical) pretest of our study.

Fig. 2: Sequence 1–3 (MR view)

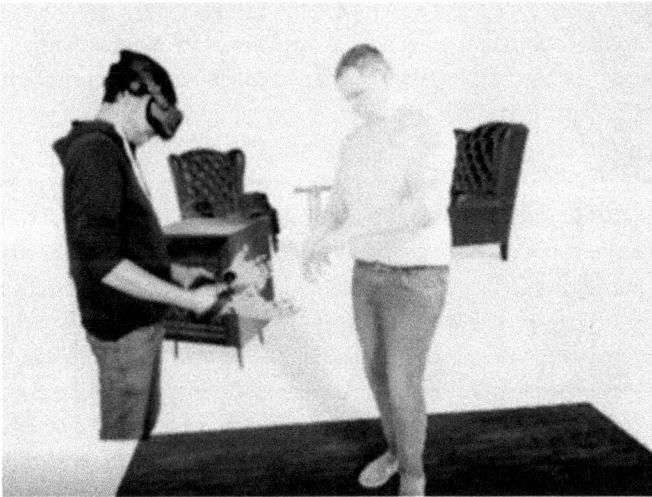

Table 2: Sequence of interactive events (1–3)

User Speech	„Hi"		
User Gaze	agent's head	user's hands	agent
User Movement			
Agent Speech			„Hello, my name is Florian. What is your name?"
Agent Gaze	user	agent's hands	user

User: The first sequences are characterized by a mutual perception of presence. The user starts by initiating eye contact and gazing at the agent's face, which is then complemented by the initiation of verbal exchange (with the utterance of "Hi"). Such behavior is consistent with the expectations of common face-to-face meetings: Mutual verification of the perception of the other's presence is usually done through glances. Additionally, the verification of presence by mirroring the agent's behavior (both the agent and the user looking at their hands) initially suggests a symmetrical situation in which both actors have similar skills and abilities for "closing" contingency. Shifting the gaze to the agent when the agent begins to speak also suggests routinized conventions of tact, respect, and politeness in conversations.

Agent: On the part of the agent, too, the perception of mutual presence by means of glances is initially striking. However, the agent does not initiate with a verbal greeting (with our external view of the agent's practical abilities, we know that the agent will act responsively here in any case). In this context, the agent's glance at his own hands, which occurs as an immediate reaction to being addressed (with "Hi"), can be interpreted in different ways: as a gesture of uncertainty or embarrassment but also as a gesture of disinterest. The fact that the agent does not directly respond with a linguistic reaction but rather seems preoccupied with himself could already be interpreted as communicating something. Finally, the agent's spoken response leaves the realm of potential rudeness. Another interpretation of this reaction to the greeting could also be understood as a difference in processing times; that is, the agent needs a longer processing period to select a reaction. However, this cannot be determined unambiguously at this point since it would take more situations of "Loading. Please wait" to align expectations with this.

Interaction: At the beginning, the interaction is (as in common human–human interactions between strangers) characterized by restraint, probing, and mutual rituals of perception. The relationship level between the two actors is still undefined. Only the agent's unusually slow reaction to the user's greeting and the gestural reaction of looking at his own hands can be considered indicators of a deviant framing. The user's symmetrizing-synchronizing gesture (looking at his hands as well), however, repairs this deviation effect.

Fig. 3: Sequence 4–6 (MR view)

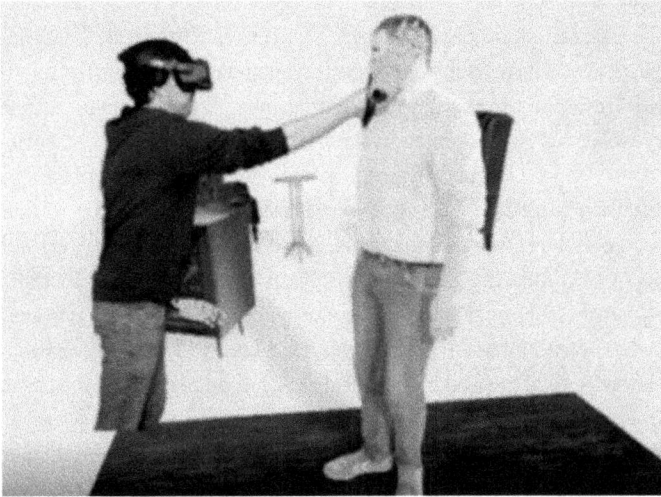

Table 3: Sequence of interactive events (4–6)

Sequence	04	05	06
User Speech	„My name is John Doe"		
User Gaze	agent's head	agent's head	agent
User Movement		„caressing/petting the agent's head"	
Agent Speech			„Nice to meet you, John Doe."
Agent Gaze	user	user	user

User: The user's answer—giving a name—accords with expectations, but the name is a pseudonym: John Doe is known to be the go-to placeholder in situations where anonymity is to be preserved. Here, an untruth is already indicated on the content level because the person can clearly be identified as the author of this paper.

The mention of this untruth is intensified shortly after by another transgression when the user pats the virtual agent's head. Such a gesture would be a clear violation of ordinary, routinized expectations in situations where strangers are getting to know each other. It must be said here, however, that the two actors in this scene are not meeting for the first time as they are both part of the research project. The head patting can thus be interpreted as a testing demonstration, a provocation aimed at eliciting any reaction. This demonstration can be interpreted as demonstrative of the agent's incapacity and the two actors' asymmetrical statuses. On a moral level, however, one would have to speak of a clear transgression of boundaries.

Agent: From the agent's side, the situation looks completely different. The boundary violations just observed in the user's behavior are not perceptible to the agent. On the one hand, the transgressive behavior therefore successfully demonstrates exactly what it was aiming at: the agent's inability to react to these kinds of actions. Looking at Suchmann's scheme, we know that from a technical viewpoint, the agent cannot integrate the user's hand movements into its repertoire of reactions, but this also means that they simply do not occur in the agent's world, which is why they would not appear in this reference as a violation of rules. Without external knowledge of the agent's practical abilities, on the other hand, it would still remain open whether the agent actually did not perceive this provocative gesture or whether it merely did not react to it. In the latter case, one would then have to ask what attitude would be behind such behavior. The purely linguistic connection ("Nice to meet you, John Doe.") would not provide any clue here, since it would still be indistinguishable whether it is a demonstration of incompetence or rather a stoic attitude characteristic of lower status persons (e.g., slaves, British royal guards, or servants).

Interaction: Overall, what was indicated in the first three sequences has now become even clearer. The interaction is characterized by an asymmetrical and hierarchizing inequality both on the relational level, which is primarily shaped by nonverbal gestures, and on the informational level, which opens up in spoken language. The user not only tests the agent's capabilities but also tries to demonstrate the agent's incompetence in multimodal comprehension. The question now increasingly arises as to what orientation might lie behind this behavior on the part of the user: Is it a pejorative attitude intended to demonstrate the agent's shortcomings through humiliation, or is it an attitude that seeks to probe the system's limits with constructive intent (much like many practical experiments in the so-called Turing Tests; cf. Turing 1950; Humphreys 2009). That being said, the gesture's multidimensionality is exposed here.

Fig. 4: Sequence 7–9 (MR view)

Table 4: Sequence of interactive events (7–9)

Sequence	07	08	09
User Speech			
User Gaze	agent	agent	agent
User Movement	„moving around the agent"	„moving around the agent"	„moving back around the agent"
Agent Speech		„What can I do for you?"	
Agent Gaze	user	user	user

User: In these last three sequences, the user makes no further verbal utterances. Rather, he moves around the agent without losing sight of the agent's face. Such movement expresses something different than if the user only moved past the agent. The gesture thus retains its observing, skeptical, and examining character.

Agent: The agent follows the user's movement with his gaze. Thus, the mutual perception of presence is uninterrupted. The sequences end with the agent asking what he can do for the user.

Interaction: From the viewpoint of interaction, the scenes here retain their testing character, which is strongly shaped by the guidance the user gives. The asymmetry in terms of both status and relationship remains clear to the end: The sequence concludes with the agent asking what he can do for the user. The agent thus remains in a servile, reactive stance, and it should now be clear that this stance will not change. The relationship between the two agents has become entrenched. At the same time, the uninterrupted turning towards each other and mutual perception indicate a high (mutual?) interest in one another.

5. Conclusions and Outlook

In summary, this paper has described the theoretical background and methodological procedures of analyzing interactions between human users and virtual agents by means of mixed reality methods. This approach aims at using mixed reality videography to collect data on the joint behavior of human users and virtual agents in a virtual environment. The methodology can then be adapted to a broad variety of research questions and can even be further extended using data from first person perspectives (i.e., user experiences).

Currently, the biggest limitation to this approach is the lack of testing experience and validation due to the effects of COVID-19 in Germany. Thus, the primary focus of this paper has been to outline the methodological approach. The main advantage of this methodology is its ability to capture interactions in VR that take place verbally and nonverbally. Using mixed reality methods, we can capture the *jointly coordinated behavior of human users and virtual agents in virtual spaces*. In contrast to just verbalizing what is happening, this approach provides a more complete picture of the richness of HAI. The mixed methods approach thus allows for the analysis of several datasets: First, the verbal interaction with the agent produces conversation protocols that can be examined in more detail by means of conversation analysis; and second, the videographic approach facilitates the examination of facial expressions, gestures, and other nonverbal communication and provides data on individual behavior.

Furthermore, this approach emphasizes the theoretical framing of interaction as a process of co-construction. For further developments of virtual humans, this aspect will likely become increasingly important because, as Kopp and Krämer recently argued,

> the basis (and also a linchpin) for the required degree of "interaction intelligence" are coordinative mechanisms such as partner-specific adaptation of multimodal utterances, responsive turn-taking, informative feedback, or collaboratively resolving misunderstandings.[62]

---

62   Kopp/Krämer, *Revisiting Human-Agent Communication.*

For this purpose, the rather unspecific talk of "interaction" between humans and technology should be extended or even replaced by the analysis of relational patterns. From this perspective, relational patterns form interfaces in which social coordination of heterogeneous elements, such as organic-physical, machine-electronic, and symbolic-cultural elements, takes place. Whereas the knowledge of HCI is highly differentiated regarding topics in engineering, current research has only scratched the surface of the phenomenological understanding of embodied and situated interactions.[63] The differentiation between human users and agents seems to be still too fixed on these two identities and the seemingly unambiguous difference between them. With the help of a relational approach in theory and methodology,[64] the corresponding ontological view on "human" and "technology" dissolves into a network of identities. From this relational sociology, it is clear that identities such as "humans" and "technologies" are shaped only within particular relations and that it is not dyads but rather networks that emerge in this process.

Viewed from this perspective, it is all about the recognition of relational patterns,[65] which describe the process in the course of which identities emerge. Consequently, a relational sociology[66] does not simply start with relations or definitions of relations but tries to explain the empirical constitution of the relations themselves. If units of inquiry are predetermined and treated in the further course as if clear from the outset, which is the case without paying attention to their varying embeddedness, then the problem that characterizes relational sociology is ignored. If, for example, man and technology, nature and society, or subject and object are treated as separate entities acting on each other, the history of their entanglements that produces them in the first place disappears.

This, in turn, brings up the question of what actually makes a good interaction. Still, it is far from clear what "quality of interaction" could actually mean.[67] However, our approach allows for the consideration of some aspects of interaction that provide information about whether we are dealing with a successful, that is, dialogic, non-faltering interaction that does not need to be constantly repaired, verified, or ratified on the basis of external factors. For example, we could assess whether both actors (human/agent) stay in the same framing. Do they talk *with* each other or *about* each other? Do they treat each other as equals or asymmetrically as in a relationship

63   Steve Harrison/Phoebe Sengers/Deborah Tatar, Making epistemological trouble: Third-paradigm HCI as successor science, in: *Interacting With Computers* 23 (2011), 385–392.

64   Werner Vogd/Jonathan Harth, Relational Phenomenology. Individual experiences and social meaning in Buddhist meditation, in: *Journal of Consciousness Studies* 26 (7–8/2019), 238–267.

65   Athanasios Karafillidis, Relationsmustererkennung. Relationale Soziologie und die Ontogenese von Identitäten, in: *Berliner Debatte Initial* 29 (4/2018), 105–125.

66   Mustafa Emirbayer, Manifesto for a Relational Sociology, in: *American Journal of Sociology* 103 (2/1997), 281–317.

67   Deriu et al., Survey on Evaluation Methods for Dialogue Systems.

that resembles that of a parent and a child, a caregiver and a person with special needs, or even a pet owner and a pet?

At the same time, however, this perspective also opens up the necessary neutral ground for establishing a serious and consistent communication theory foundation for the conditions of interaction between human and nonhuman interaction partners.[68] After all, a culturally regimented interaction with more or less different beings would truly be nothing new. In history, we have always found ways of dealing not only with other selves but also with other others. In dealing with cultural strangers, mentally impaired people, small children or babies, animals, and even ghosts or gods, very specific social relational patterns have emerged in each case. To add technical or other artificial nonhuman entities to this list would only be a small step.

Nevertheless, the data must be considered in the light of at least two important limitations: First, ECAs are far from being part of the everyday life experience; they are constantly evolving in terms of both technology and variety. Social norms for dealing with artificially intelligent entities have still not emerged in the mainstream. Therefore, interactions with chatbots, ECAs, and virtual humans are still of an exploratory and experimental nature. The interaction with our prototype is thus characterized by novelty (or strangeness) from the viewpoint of the subject matter alone. Second, the setting in which this interaction takes place is characterized by novelty as well: Visiting a university lab and participating in an experiment frame the interaction as extraordinary and unique. Moreover, for some users, it may even be their first time immersing in VR. Thus, we must consider that both the object and the situation are characterized by a high degree of novelty, unusualness, and ambiguity. For the users, there will be the question of how to behave towards the virtual agent, as well as the question of how to present themselves in front of the camera. Consequently, the situation in the research lab has to be understood as unusual in at least two respects. These contexts will generate corresponding expectations for the subjects (e.g., social desirability).

Beyond that, further technical elaboration of virtual agents is much needed. Virtual agents must possess the ability to process nonverbal inputs from users. Only then can we speak of an equally structured two-way interaction that does not pause at the illegibility of human nonverbal signals. On the level of verbal exchange, the participants and the agent are on symmetrical ground, but with regard to nonverbal

---

68   See Jonathan Harth, Empathy with non-player characters? An empirical approach to the foundations of human/non-human-relationships, in: *Journal of Virtual Worlds Research* 10 (2/2017) and Jonathan Harth, Simulation, Emulation oder Kommunikation? Soziologische Überlegungen zu Kommunikation mit nicht-menschlichen Entitäten In: Schetsche, Michael/Anton, Andreas (eds.), *Intersoziologie. Menschliche und nichtmenschliche Akteure in der Sozialwelt*, Weinheim 2021, 143–158.

expression, only the user can interpret the potential meaning of the agent's behavior. What humans succeed in doing naturally, namely bringing about a communicative action multimodally, is still a major challenge for virtual agents and humanoid robots.[69]

However, current technological research and development is heading in precisely this direction: Recently, DeepMind announced their new research program whose goal is "to build embodied artificial agents that can perceive and manipulate the world, understand and produce language, and react capably when given general requests and instructions by humans."[70] Facebook as well is fully committed to developing both situated and multimodal agents for rich HAI.[71] Together with evolving machine learning algorithms, which are domain-agnostic and able to generalize to new environments not seen during training,[72] future research on HAI seems more than bright. These foreseeable technical advancements will require powerful research methods as well.

Bibliography

Adiwardana, Daniel/Luong, Minh-Thang/So, David R./Hall, Jamie/Fiedel, Noah/ Thoppilan, Romal/Yang, Zi/Kulshreshtha, Apoorv/Nemade, Gaurav/Lu, Yifeng/Le, Quoc V., Towards a Human-like Open-Domain Chatbot, in: arXiv:2001.09977, 2020.

Alexandrovsky, Dmitry/Putze, Susanne/Schwind, Valentin/Mekler, Elisa D./ Smeddinck, Jan D./Kahl, Denise/Krüger, Antonio/Malaka, Rainer Evaluationg User Experiences in Mixed Reality, in: arXiv preprint arXiv:2101.06444, 2021.

Bartneck, Christoph/Belpaeime, Tony/Eyssel, Friederike/Kanda, Takayuki/Keijsers, Merel/Sabanovic, Selma, *Human-Robot Interaction. An Introduction*. Cambridge 2019.

Bavelas, Janet B./Hutchinson, Sarah/Kenwood, Christine Matheson, Deborah H., Using face-to-face dialogue as a standard for other communication systems, in: *Canadian Journal of Communication* 22 (1/1997), 5.

Bischof, Andreas, *Soziale Maschinen bauen. Epistemische Praktiken der Sozialrobotik*, Bielefeld 2017.

---

69   Nadia Magnenat Thalmann et al., *Context Aware Human-Robot and Human-Agent Interaction*, Singapore 2016.

70   DeepMind Interactive Agents Group, *Imitating Interactive Intelligence*, in: arXiv:2012.05672v2, 2021.

71   Seungwhan Moon et al., *Situated and Interactive Multimodal Conversations*, URL: https://githu b.com/facebookresearch/simmc [last accessed: August 15, 2023].

72   John D. Co-Reyes et al., *Evolving Reinforcement Learning Algorithms*, in: arXiv:2101.03958v3 [cs.LG] [last accessed: August 15, 2023].

Bock, Indra/Mayer, Henning, Humanoide Roboter und virtuelle Agenten als Kommunikationsteilnehmer? Konversationsanalytische Studien der Mensch-Maschine-Interaktion, in: Helen Ahner/Max Metzger/Mathis Nolte (eds.), *Von Menschen und Maschinen: Interdisziplinäre Perspektiven auf das Verhältnis von Gesellschaft und Technik in Vergangenheit, Gegenwart und Zukunft; Proceedings der 3. Tagung des Nachwuchsnetzwerks „INSIST", 05.–07. Oktober 2018, Karlsruhe*, Karlsruhe 2020.

Breazeal, Cynthia, Toward sociable robots, in: *Robotics and Autonomous Systems* 42 (3–4/2003), 167–175.

Burden, David/Savin-Baden, Maggi, *Virtual Humans. Today and Tomorrow*, Boca Raton 2020.

Cassel, Justine/Sullivan, Joseph W./Prevost, Scott/Churchill, Elisabeth F., *Embodied Conversational Agents*, Cambridge; Massachusetts; London 2000.

Co-Reyes, John D./Miao, Yingjie/Peng, Daiyi/Real, Esteban Levine, Sergey/Le, Quoc V./Lee, Honglak/Faust, Aleksandra, *Evolving Reinforcement Learning Algorithms*, in: arXiv:2101.03958v3 [cs.LG] [last accessed: May 10, 2021].

DeepMind Interactive Agents Group, *Imitating Interactive Intelligence*, in: arXiv:2012.05672v2, 2021.

Deriu, Jan A./Arantxa, Rodrigo/Echegoyen, Otegi G./Rosset, Sophie/Agirre, Eneko/ Cieliebak, Mark, Survey on Evaluation Methods for Dialogue Systems, in: *arXiv*:1905.04071v2, 2020.

Digital Humans Inc., URL: https://digitalhumans.com/ [last accessed: August 15, 2023].

Digital Domain, *Introducing Douglas—Autonomous Digital Human*, URL: https://www.youtube.com/watch?v=RKiGfGQxqaQ [last accessed: August 15, 2023].

Dingemanse, Mark/Roberts, Seán G./Baranova, Julija/Blythe, Joe/Drew, Paul/ Floyd, Simeon/Gisladottir, Rosa S./Kendrick, Kobin H./Levinson, Stephen C./ Manrique, Elizabeth/Rossi, Giovanni/Enfield, N. J., Universal Principles in the Repair of Communication Problems, in: *PLOS ONE* 2015, e0136100.

Doyle, Philip R./Edwards, Justin/Dumbleton, Odile/Clark, Leigh/Cowan, Benjamin R., Mapping Perceptions of Humanness in Intelligent Personal Assistant Interaction, in: *Proceedings of the 21st International Conference on Human-Computer Interaction with Mobile Devices and Services*, 2019.

Emirbayer, Mustafa, Manifesto for a Relational Sociology, in: *American Journal of Sociology* 103 (2/1997), 281–317.

Fitrianie, Siska/Bruijnes, Merijn/Richards, Deborah/Abdulrahman, Amal/ Brinkman, Willem-Paul, What are We Measuring Anyway? A Literature Survey of Questionnaires Used in Studies Reported in the Intelligent Virtual Agent Conferences, in: *Proceedings of the 19th ACM International Conference on Intelligent Virtual Agents*, 2019, 159–161.

Fitrianie, Siska/Bruijnes, Merijn/Richards, Deborah/Bönsch, Andrea/Brinkman, Willem-Paul, The 19 Unifying Questionnaire Constructs of Artificial Social

Agents: An IVA Community Analysis, in: *Proceedings of the 20th ACM International Conference on Intelligent Virtual Agents (IVA '20)*. Association for Computing Machinery, New York 2020, Article 21, 1–8.

Foster, Mary E., Face-to-Face Conversation: Why Embodiment Matters for Conversational User Interfaces, in: *Proceedings of the 1st International Conference on Conversational User Interfaces (CUI '19)*. Association for Computing Machinery, New York 2019, Article 13, 1–3.

Freigang, Farina/Klett, Sören/Kopp, Stefan, Pragmatic Multimodality: Effects of Nonverbal Cues of Focus and Certainty in a Virtual Human, in: Jonas Beskow/Christopher Peters/Ginevra Castellano/Carol O'Sullivan/ Iolanda Leite/Stefan Kopp (eds.), *Intelligent Virtual Agents*. 17th International Conference, IVA 2017, Stockholm, Sweden, August 27–30, 2017, Proceedings, 142–155.

Garfinkel, Harold, *Studies in Ethnomethodology*, Englewood Cliffs 1967.

Garfinkel, Harold, *Ethnomethodology's Program. Working out Durkheim's Aphorism*, Lanham 2002.

Goffman, Erving, *Encounters. Two Studies in the Sociology of Interaction*, Hamrondsworth 1972.

Gong, Li, How social is social responses to computers? The function of the degree of anthropomorphism in computer representations, in: *Computers in Human Behavior* 24 (4/2008), 1494–1509.

Harrison, Steve/Sengers, Phoebe/Tatar, Deborah, Making epistemological trouble: Third-paradigm HCI as successor science, in: *Interacting With Computers* 23 (2011) 385–392.

Harth, Jonathan, Empathy with non-player characters? An empirical approach to the foundations of human/non-human-relationships, in: *Journal of Virtual Worlds Research* 10 (2/2017).

Harth, Jonathan, Simulation, Emulation oder Kommunikation? Soziologische Überlegungen zu Kommunikation mit nicht-menschlichen Entitäten, in: Michael Schetsche/Andreas Anton (eds.): *Intersoziologie. Menschliche und nichtmenschliche Akteure in der Sozialwelt*, Weinheim 2021, 143–158.

Harth, Jonathan/Feißt, Martin, Neue soziale Kontingenzmaschinen. Überlegungen zu künstlicher sozialer Intelligenz am Beispiel der Interaktion mit GPT-3, in: Martin Schnell/Lukas Nehlsen (eds.): *Begegnungen mit künstlicher Intelligenz. Intersubjektivität, Technik, Lebenswelt*, Velbrück 2022.

Hayashi, Makoto/Raymond, Geoffrey/Sidnell, Jack (eds.), *Conversational Repair and Human Understanding*, Cambridge 1992.

Heritage, John, Conversation Analysis and Institutional Talk: Analyzing Distinctive Turn-Taking Systems, in: Svetla Cmejrková, Jana Hoffmannová, Olga Müllerová and Jindra Svetlá (eds.), *Proceedings of the 6th International Congresss of IADA (International Association for Dialog Analysis)*, Tübingen 1998, 3–17.

Huang, Hung-Hsuan, Embodied conversational agents, in: Kent L. Norman/Jurek Kirakowski (eds.), *The Wiley handbook of human computer interaction*, Blackwell 2018, 601–614.

Karafillidis, Athanasios, Relationsmustererkennung. Relationale Soziologie und die Ontogenese von Identitäten, in: *Berliner Debatte Initial* 29 (4/2018), 105–125.

Knoblauch, Hubert/Tuma, René, Videography: an interpretive approach to video-recorded micro-social interaction, in: Eric Margolis/Luc Pauwels (eds.), *The Sage Handbook of Visual Methods*, Thousand Oaks 2011, 414–430.

Kopp, Stefan/Krämer, Nicole, Revisiting Human-Agent Communication: The Importance of Joint Co-construction and Understanding Mental States, in: *Frontiers of Psychology* 12 (2021), 580955.

Krummheuer, Antonia, Conversation Analysis, Video Recordings, and Human-Computer Interchanges, in: Ulrike T. Kissmann (ed.), *Video Interaction Analysis. Methods and Methodology*, Frankfurt a. M. 2009, 59–83.

Krummheuer, Antonia, *Interaktion mit virtuellen Agenten? Zur Aneignung eines ungewohnten Artefakts*, Stuttgart 2010.

LIV, URL: https://liv.tv/ [last accessed: August 15, 2023].

Lowe, Ryan/Noseworthy, Michael/Serban, Iulian V./Angelard-Gontier, Nicolas/Bengio, Yoshua/Pineau, Joelle, Towards an automatic turing test: Learning to evaluate dialogue responses, in: arXiv:1708.07149v2, 2018.

Magnenat-Thalmann, Nadia/Yuan, Junsong/Thalmann, Daniel/You, Blum-Jae, *Context Aware Human-Robot and Human-Agent Interaction*. Singapore 2016.

Mayer, Henning/Muhle, Florian/Bock, Indra, Whiteboxing MAX. Zur äußeren und inneren Interaktionsarchitektur eines virtuellen Agenten, in: Eckhard Geitz/Christian Vater/Silke Zimmer (eds.), *Black Boxes-Versiegelungskontexte und Öffnungsversuche*, Berlin 2020, 295–322.

Meadan, Hedda/Halle, James W., Communication Repair and Response Classes, in: *The Behavior Analyst Today* 5 (3/2004).

Miao, Fred/Kozlenkova, Irina V./Wang, Haizhong/Xie, Tao/Palmatier, Robert W., EXPRESS: An Emerging Theory of Avatar Marketing, in: *Journal of Marketing* (4/2021).

Magic Leap Inc, *Magic Leap's Mica at GDC*, URL: https://www.youtube.com/watch?v=-PzeWxtOGzQ [last accessed: August 15, 2023].

Mondada, Lorenza, Conversation Analysis and Institutional Interaction, in: *The Encyclopedia of Applied Linguistics* (2012).

Moon, Seungwhan/Kottur, Satwik/Crooky, Paul A./Dey, Ankita/Poddary, Shivani/Levin, Theodore/Whitney, David/Difranco, Daniel/Beirami, Ahmad/Cho, Eunjoon/Subba, Rajen/Geramifard, Alborz, *Situated and Interactive Multimodal Conversations*, URL: https://github.com/facebookresearch/simmc [last accessed: August 15, 2023].

Mori, Masahiro/MacDorman, Karl F./Kageki, Norri, The Uncanny Valley, in: *IEEE Robotics &Automation Magazine* 19 (2012), 98–100.

Muhle, Florian, Sozialität von und mit Robotern? Drei soziologische Antworten und eine kommunikationstheoretische Alternative, in: *Zeitschrift für Soziologie* 47 (3/2018), 147–163. See also Florian Muhle, Humanoide Roboter als 'technische Adressen'. Zur Rekonstruktion einer Mensch-Roboter-Begegnung im Museum, in: *Sozialer Sinn* 20 (1/2019), 85–128.

Muhle, Florian, Humanoide Roboter als 'technische Adressen'. Zur Rekonstruktion einer Mensch-Roboter-Begegnung im Museum, in: *Sozialer Sinn* 20 (1/2019), 85–128.

Reinhardt, Jens/Hillen, Luca/Wolf, Katrin, Embedding Conversational Agents into AR: Invisible or with a Realistic Human Body?, in: *Proceedings of the Fourteenth International Conference on Tangible, Embedded, and Embodied Interaction*, ACM Press, New York 2020.

Sacks, Harvey/Schegloff, Emanuel A./Jefferson, Gail, A Simplest Systematics for the Organization of Turn-Taking for Conversation, in: *Language* 50 (4/1974), 696–735.

Sacks, Harvey, Notes on Methodology, in: J. Maxwell Atkinson/John Heritage (eds.), *Structures of Social Actions. Studies in Conversation Analysis*, Cambridge 1984, 21–27.

Schmitt, Reinhold, Positionspapier: Multimodale Interaktionsanalyse, in: Ulrich Dausendschön-Gay/Elisabeth Gülich/Ulrich Krafft (eds.), *Ko-Konstruktionen in der Interaktion: Die gemeinsame Arbeit an Äußerungen und anderen sozialen Ereignissen*, Bielefeld 2015, 43–51.

Seymour, Mike/Evans, Chris/Libreri, Kim, Meet Mike: epic avatars, in: *ACM SIGGRAPH 2017 VR Village (SIGGRAPH '17)*. Association for Computing Machinery, New York, NY, USA, Article 12, (2017), 1–2.

Sidnell, Jack, *Conversation analysis. An introduction*, Chichester 2010.

SoulMachines, URL: https://www.soulmachines.com/ [last accessed: August 15, 2023].

Suchman, Lucy, *Plans and situated actions: The problem of human-machine communication*, Cambridge 1987.

Suchman, Lucy, *Human-Machine Reconfigurations. Plans and Situated Actions* (2nd Edition), Cambridge 2007.

Thaler, Markus/Schlögl, Stephan/Groth, Aleksander, Agent vs. Avatar: Comparing Embodied Conversational Agents Concerning Characteristics of the Uncanny Valley, in: *IEEE International Conference on Human-Machine Systems (ICHMS)*, 2020.

Vogd, Werner/Harth, Jonathan, Relational Phenomenology. Individual experiences and social meaning in Buddhist meditation, in: *Journal of Consciousness Studies* 26 (7–8/2019), 238–267.

Watzlawick, Paul/Beavin, Janet H./Jackson, Doti D., *Pragmatics of Human Communication. A Study of Interactional Patterns, Pathologies, and Paradoxes*, New York 1967.

Watzlawick, Paul/Beavin, Janet, Some Formal Aspects of Communication, in: *The American Behavioral Scientist* 10 (8/1967), 4–8.

Weinstein, James N., Artificial Intelligence: Have You Met Your New Friends; Siri, Cortana, Alexa, Dot, Spot, and Puck, in: *Spine* 44 (1/2019), 1–4.

Weiss, Benjamin/Wechsung, Ina/Kühnel, Christine/Möller, Sebastian, Evaluating embodied conversational agents in multimodal interfaces, in: *Computational Cognitive Science* 1 (6/2015).

Problems and Possibilities of Interaction With MAX
Investigating the Architecture-for-Interaction
of an Embodied Conversational Agent in a Museum

Florian Muhle, Indra Bock, Henning Mayer

Abstract *Despite technical progress in the development of communicative artificial intelligence, problems and difficulties still occur regularly in human–machine communication. These can be attributed, in part, to problems in the interface design but also to limitations in technical systems' communicative capabilities. However, how exactly system design and programming contribute to the emergence of communicative problems in human–machine communication has rarely been studied so far. Against this background, this chapter presents an analytical approach that systematically interweaves the analysis of communicative technical system design and programming with the empirical study of human–machine communication. This approach is inspired by the analytical concept of "architectures-for-interaction," which was developed in recent years through linguistic research. After introducing the approach, its analytical potential will be illustrated by examining an encounter between an embodied agent and a human user in a computer museum.*

1. Introduction

For a long time, human–machine communication was tied to the instrumental operation of computers, which were used primarily as tools. For some years now, however, this situation has begun to change due to the development of "communicative AI (artificial intelligence)."[1] Thanks to the establishment of artificial interlocutors in the form of robots, smart speakers, or virtual agents, the idea of humanlike interaction with technical objects is no longer a vision of the future that is only realized in science fiction movies. Instead, interaction with (more or less) humanlike machines now takes place—at least in a rudimentary way—in everyday life. Smart speakers that can be operated via voice control can be found in private living rooms,

---

1   Andrea L. Guzman/Seth C. Lewis, Artificial Intelligence and Communication: A Human–Machine Communication Research Agenda, in: *New Media & Society* 22 (1/2020), 70–86.

and robots and agents are used, at least for testing purposes, in institutional contexts—be it in the fields of education[2] or health care,[3] or in museums, where conversational machines serve as visitor guides.[4]

However, as numerous empirical analyses have shown, current systems still
have not achieved intuitive and "natural" communication.[5] Instead, comprehension
problems regularly occur in human–machine interactions.[6] Additionally, users
sometimes have difficulties contacting their machine counterparts since the possibilities and limits of interacting with the technical systems are not immediately
apparent.[7]

To get to the bottom of these problems and understand in detail how and
why they appear regularly in human–machine encounters, qualitative research
approaches to human–machine communication are particularly suitable since
they are characterized by special proximity to their research object and context
sensitivity. Accordingly, in this chapter, we would like to present and apply such
an approach. Its distinctive feature is that it systematically intertwines analysis
of problems in human–machine communication with consideration of the design

---

2   Omar Mubin et al., A Review of the Applicability of Robots in Education, in: *Technology for Education and Learning* 1 (1/2013); Tony Belpaeme et al., Social robots for education: A review, in: *Science robotics* 3 (21/2018).

3   Moojan Ghafurian/Jesse Hoey/Kerstin Dautenhahn, Social Robots for the Care of Persons with Dementia, in: *ACM Trans. Hum.-Robot Interact.* 10 (4/2021), 1–31; Jane Holland et al. Service Robots in the Healthcare Sector, in: *Robotics* 10 (1/2021), 47.

4   Timothy Bickmore/Laura Pfeifer/Daniel Schulman, Relational Agents Improve Engagement and Learning in Science Museum Visitors, Boston 2011; Karola Pitsch, Referential Practices for a Museum Guide Robot. Human-Robot-Interaction as a Methodological Tool to Investigate Multimodal Interaction, in: Mensch und Computer 2019 – Workshopband, Bonn 2019; Stefan Kopp et al., A Conversational Agent as Museum Guide – Design and Evaluation of a Real-World Application, in: Themis Panayiotopoulos/Jonathan Gratch/Ruth Aylett/Daniel Ballin/Patrick Olivier/Thomas Rist (eds.), Intelligent Virtual Agents, Springer Berlin, Heidelberg 2005, 329–343.

5   Florian Muhle/Indra Bock, Intuitive Interfaces? Interface Design and its Impact on Human-Robot Interaction, in: *Mensch und Computer 2019 – Workshopband*, Bonn 2019, 346–347; Stuart Reeves,/Martin Porcheron/Joel Fischer, "This is Not What We Wanted": Designing for Conversation with Voice Interfaces, in: *Interactions* 26 (1/2018), 46–51; Antonia Krummheuer, Zwischen den Welten. Verstehenssicherung und Problembehandlung in künstlichen Interaktionen von menschlichen Akteuren und personifizierten virtuellen Agenten, in: Herbert Willems (ed.), *Weltweite Welten. Internet-Figurationen aus wissenssoziologischer Perspektive* (1st edition), Wiesbaden 2008, 269–294.

6   Krummheuer, Zwischen den Welten; Lucy A. Suchman, *Human-machine reconfigurations. Plans and situated actions* (2nd edition), Cambridge 2007.

7   Muhle/Bock, Intuitive Interfaces?.

and programming of technical interaction partners.[8] This approach is inspired by the analytical concept of "architectures-for-interaction," which, in recent years, was developed in linguistic research interested in how "architecture enables and suggests social interaction."[9]

In what follows, we will first briefly introduce our approach to analyzing architectures-for-interaction (Section 2), which will serve as the basis to investigate a particular technical system's architecture-for-interaction: the embodied agent MAX, an exhibition object in the computer museum Heinz Nixdorf MuseumsForum (HNF) in Paderborn, Germany. In our analysis, we distinguish between the agent system's "external" and "internal" architecture, referring to MAX's visible interface design (Section 3) and invisible programming (Section 4), respectively. To show how the architecture pre-structures the interaction possibilities and limitations, the analysis is intertwined with the investigation of a short encounter between a museum visitor and MAX that provides detailed insights into understanding the possibilities and problems of human–machine interaction, while simultaneously making a relevant contribution to the analysis of technical artifacts.

2. Architectures-for-Interaction

The concept of architectures-for-interaction has emerged in recent years in the context of ethnomethodologically inspired linguistic research, which—like other approaches in the wake of material, practice, and spatial turns—is increasingly interested in the significance of materiality, space, and embodiment for the social production of meaning.[10] Compared to other (materialist) approaches,[11] the special feature of the architecture-for-interaction concept is that, firstly, it combines consideration of the material environment with an interest in investigating interactions and,

---

8   Henning Mayer/Florian Muhle/Indra Bock, Whiteboxing MAX. Zur äußeren und inneren Interaktionsarchitektur eines virtuellen Agenten, in: Eckhard Geitz/Christian Vater/Silke Zimmer-Merkle (eds.), *Black Boxes—Versiegelungskontexte und Öffnungsversuche. Interdisziplinäre Perspektiven* (1st edition), Berlin 2020.

9   Heiko Hausendorf/Reinhold Schmitt, Architecture-for-interaction: Built, designed and furnished space for communicative purposes, in: Andreas H. Jucker/Heiko Hausendorf (eds.), *Pragmatics of Space*, Berlin, Boston 2022, 431–472, see 441.

10   Hausendorf/Schmitt, Architecture-for-interaction; Heiko Hausendorf/Reinhold Schmitt/Wolfgang Kesselheim (eds.), *Interaktionsarchitektur, Sozialtopographie und Interaktionsraum*, Tübingen 2016.

11   Bruno Latour, Reassembling the social. An introduction to actor network theory, Oxford 2005; Carl Knappett, /Lambros Malafouris (eds.), Material agency. Towards a non-anthropocentric approach, New York 2008.

secondly, in the tradition of ethnomethodology, it attaches importance to developing empirical insights "from the data themselves."[12] On the one hand, this means that material phenomena are not of interest as such—only with regard to how they (a) afford certain forms of interaction and (b) are themselves made relevant in the course of interaction processes. On the other hand, the data-driven analysis perspective goes hand in hand with the fact that the phenomena of interest are not approached in a category-guided way but rather in an open-ended manner, yielding detailed and deep insights into their characteristics.

The core idea of the analysis of architectures-for-interaction is to interpret spatial and material arrangements as "architecturally manifested social expectation[s]"[13] that suggest certain movements, perceptions, and (inter-)actions among those who are (co-)present in a certain space.[14] In this way, material arrangements and the design of materials and objects can be used by (inter)actors as resources for situated sense-making. This is possible since architectural arrangements provide "usability cues" as "built-in spatial features that allow for certain forms of use and, moreover, suggest not only possible, but rather the more probable and most likely forms of use."[15]

Hausendorf and Schmitt have distinguished three types of usability cues: basic navigational cues, acquired reading cues, and full-fledged participation cues.[16] Whereas basic navigational cues provide basic information on how to navigate in and through a particular space and are related "to users' basic perceptual and motor skills,"[17] reading cues have a prerequisite: They require some kind of "architectural literacy"[18] and thus a certain know-how in terms of knowing how to open a door, use a chair, or open a window. Even more elaborate are participation cues since they

> call for understanding in a deeper sense. Institutional architectures ("churches," "hospitals," "university buildings," "court rooms," etc.) are abuzz with participation cues of this kind, so that situating oneself in such a space already implies social positioning in terms of rights and duties.[19]

---

12   Heiko Hausendorf/Reinhold Schmitt, Standbildanalyse als Interaktionsanalyse: Implikationen und Perspektiven, in: Heiko Hausendorf/Reinhold Schmitt/Wolfgang Kesselheim (eds.), *Interaktionsarchitektur, Sozialtopographie und Interaktionsraum*, Tübingen 2016, 161–187., see 166; Emanuel A. Schegloff/Harvey Sacks, Opening up closings, in: *Semiotica* 8 (4/1973), 289–327, see 291.

13   Hausendorf/Schmitt, Architecture-for-interaction, 450.

14   Hausendorf/Schmitt, Architecture-for-interaction, 442.

15   Hausendorf/Schmitt, Architecture-for-interaction, 442.

16   Hausendorf/Schmitt, Architecture-for-interaction, 442.

17   Hausendorf/Schmitt, Architecture-for-interaction, 442.

18   Hausendorf/Schmitt, Architecture-for-interaction, 442.

19   Hausendorf/Schmitt, Architecture-for-interaction, 443.

In this sense, participation cues "give hints [...] of a certain communicative framework that relates to participation in a certain social practice."[20] For instance, Hausendorf and Schmitt have shown how the architecture of a university lecture hall allows students to take their place in the lecture hall's rows of seating in a particular way and thus position themselves "as part of the audience and public that [are] an essential part of the expected social practice (the 'lecture')."[21]

However, the decisive factor here is that the cues do not determine the next actions but rather open possibility spaces for (communicative) activities. That is, they enable and suggest particular forms of "social interaction, albeit without having the ability to determine or forestall what will take place."[22] Students' occasional occupations of lecture halls, for example, indicate that the lecture halls can also be used for purposes other than lectures and can even offer opportunities for staying overnight. Therefore, it is important to carry out empirical analyses in a fine-grained manner to develop different readings of the architecture of interest and thus to reconstruct the full range of "architectural implications for social interaction"[23] and not just the most obvious ones.

Hausendorf and Schmitt have proposed carrying out the analysis of architectures-for-interaction as the first part of a comprehensive interaction analysis. For this, they suggest using still images of video recordings (without interacting persons, if possible) or other data such as photos, drawings, plans, or floor plans of the architectural phenomena of interest.[24] When analyzing these documents, the first task is to examine interactional implications that can be found simultaneously in the object under investigation. This part of the reconstruction accordingly aims at shedding light on the possible manifold of expectations that relate to particular architectural manifestations. Based on this, in the second step, the interactional relevance of the architecture is to be examined in its sequential position as a starting point for interaction. The aim is to clarify what can be expected as a next action according to the reconstructed architecturally manifested social expectations.

The analysis of architecture-for-interaction and its social implications serve as the basis for the investigation of activities that actually take place within the given material setting. This investigation can be treated as the third step of the analysis. It not only shows how people actually behave but also facilitates the relation of this behavior back to the previously reconstructed implications of the architecture. In this way, this step yields insights into how exactly architectural implications become relevant in and for interaction.

---

20   Hausendorf/Schmitt, Architecture-for-interaction, 443.
21   Hausendorf/Schmitt, Architecture-for-interaction, 450.
22   Hausendorf/Schmitt, Architecture-for-interaction, 441.
23   Hausendorf/Schmitt, Architecture-for-interaction, 441.
24   Hausendorf/Schmitt, Standbildanalyse als Interaktionsanalyse, 177.

As conceived by its proponents, the concept of architecture-for-interaction refers primarily to the built and designed space in which people navigate and (inter-)act. However, we believe that it can also be fruitfully applied to the analysis of concrete technical artifacts and their interfaces, as well as to the analysis of the internal system architecture of these artifacts, namely their programming. Thus, we propose considering the concept of architecture-for-interaction as a general analytical framework that can be used for

(1) the analysis of the spatial embedding of an (technical) artifact,
(2) the analysis of an artifact's interface, and
(3) the analysis of the artifact's internal programming

in terms of how these material manifestations provide usability cues and hence enable and restrict possibilities for (inter-)action with the artifacts.

Regarding the analysis of interface design, such an adaptation is relatively easy. For this, as Hausendorf and Schmitt have also suggested, the concept of architecture-for-interaction can be related to considerations about the "affordances"[25] of communication technologies, which have been prominent for some years in the context of the sociology of technology and beyond,[26] to analyze how communication technologies' material design opens up—but also restricts—mediatized interaction possibilities.[27]

The difference between the "classical" analysis of architectures-for-interaction and the analysis of communication technologies' affordances lies only in the divergent research objects (built and designed space vs. technical artefacts). Additionally, the analysis of affordances is decidedly aimed at aspects such as usability and is thus

---

25   The concept of affordances originally stems from the psychologist James J. Gibson (James J. Gibson, The Theory of Affordances, in: Robert Shaw/John Bransford (eds.), *Perceiving, acting, and knowing. Toward an ecological psychology*, Hillsdale 1977, 67–82).

26   Ian Hutchby, Technologies, Texts and Affordances, in: *Sociology* 35 (2/2001), 441–456; Ilkka Arminen/Christian Licoppe/Anna Spagnolli, Respecifying Mediated Interaction, in: *Research on Language and Social Interaction* 49 (4/2016), 290–309; William W. Gaver, Technology affordances, in: Scott P. Robertson (ed.), *Reaching through technology*, in: *CHI '91; conference proceedings; New Orleans, Louisiana, April 27—Mai 2 1991. the SIGCHI conference*, New Orleans 1991, 79–84; David Martin/John Bowers/David Wastell, The Interactional Affordances of Technology. An Ethnography of Human-Computer Interaction in an Ambulance Control Centre, in: Harold Thimbleby/Brid O'Conaill/Peter J. Thomas (eds.), *People and Computers XII. Proceedings of HCI '97*, London 1997, 263–281; Andrew R. Schrock, Communicative affordances of mobile media: Portability, availability, locatability, and multimediality, in: *International Journal of Communication* 9 (2015), 18; Nicole Zillien, Die (Wieder-)Entdeckung der Medien – Das Affordanzkonzept in der Mediensoziologie, in: *Sociologia Internationalis* 46 (2/2008), 161–181.

27   Hausendorf/Schmitt, Architecture-for-interaction, 442.

more specific than the analysis of an artifact's spatial embedding, which also deals with questions of sit-on-ability, look-at-ability, or walk-on-ability.[28]

The situation of internal programming is somewhat more difficult. In the case of the embodied agent MAX, the developers actually also refer to the internal programming as "system architecture."[29] In contrast to the spatial embedding and the interface design, the system architecture is not directly visible to those who "interact" with the technical system. Instead, it remains a "black box" that only indirectly provides information about its functioning via the system's output. Accordingly, the system architecture does not provide any visible usability cues. Nevertheless, there is no doubt that it significantly contributes to opening up and restricting interaction possibilities, which is why it can be understood as a further component of the architecture-for-interaction of the artifact under investigation.

Analyses of technical artifacts' internal architecture-for-interaction thus complement investigations of the communicative implications of the visible aspects of the interface design and the artifact's situated embedding, thus enabling a comprehensive understanding of human–machine communication (and its problems). To carry out analyses of technical artifacts' internal architectures, however, the data required differ from those required for the analysis of interfaces and designed space. Photographs and still images do not help here. Instead, information about the system architecture and the concrete program code is needed to reconstruct "what is made possible and facilitated, and what is made difficult and inhibited"[30] by the artifact's internal functioning. Accordingly, in this respect, the analysis of architectures-for-interaction takes the form of (critical) code studies[31] dealing with the social analysis of computer code.

To show the architecture-for-interaction concept's analytical potential for the analysis of human–machine communication, we will, in what follows, conduct an exemplary analysis of the agent MAX's "external" (Section 3) and "internal" (Section 4) architecture and combine it with the empirical analysis of an actual encounter between MAX and a museum visitor. The focus will be on investigating the interface and programming. The artifact's spatial embedding is included in the analysis of the interface since both are inextricably intertwined in the present case (see Section 3).

---

28   Hausendorf/Schmitt, Architecture-for-interaction, 442.

29   Kopp et al., A Conversational Agent as Museum Guide.

30   Eve Bearne/Gunther Kress, Editorial, in: *Reading: literacy and language* 35 (3/2001), 89–93, see 91.

31   Mark C. Marino, *Critical code studies*, Cambridge, Massachusetts 2020.

3. Analysis of MAX's Interface

MAX is a so-called embodied conversational agent[32], originally developed at the University of Bielefeld "for studying the generation of natural multimodal behavior."[33] Since 2004, a version of MAX has been on permanent exhibition at the HNF, a public computer museum, where it acts as a museum guide.[34] Its main task is "to engage visitors in conversations in which he provides them[,] in comprehensible and interesting ways[,] with information about the museum, the exhibition, or other topics of interest."[35]

3.1   The Interface's Architecture

To analyze MAX's interface, we will use a photograph that shows the exhibit in its entirety and an additional two photographs that are detailed shots of the "object identifiers" belonging to the object (see Figures 1–3).

The first photograph (Fig. 1) shows how the agent is presented in the museum at a time in the morning when no one has "interacted" with him yet. It thus documents the system's "basic state" as it exists in the morning after being booted.

In accordance with the methodological considerations outlined above (Section 2), the first step of the analysis is to take a look at the simultaneously recognizable aspects of the exhibition object. When conducting this step, it is immediately noticeable that MAX, in the narrower sense, is not a single object but rather an entire "artifact arrangement", comprising:

1. a canvas on which the agent can be seen but also a two-line text field below the visible agent,
2. a bar table with an integrated keyboard,
3. a camera positioned to the right of the screen at approximately the height of an adult visitor's head,
4. a table below the canvas, and
5. two signs, one on the bar table and another on the table below the canvas.

---

32   The communicative status of the agent MAX is not clear. Sometimes MAX is treated as an object, sometimes as a subject. To reflect this ambiguity, we occasionally use the pronoun 'he', but sometimes also 'it' with reference to the agent.

33   Kopp et al., A Conversational Agent as Museum Guide, 329.

34   For a long time, MAX was part of the Artificial Intelligence and Robotics exhibition area, but since autumn 2018, he has been on display in the newly designed Human, Robot! area, and his appearance has changed slightly. However, our analysis refers to the former MAX, as we collected our data before the redesign of the exhibition area.

35   Kopp et al., A Conversational Agent as Museum Guide, 330.

Fig. 1: Presentation of the agent in the museum

The agent in its embodied form on the canvas is therefore only part of a complex interface that is composed of several artifacts. In the museum context, the two signs on the tables especially suggest that the agent system is not only a technology for interaction but also a museum exhibit. Even if the signs cannot be read from the distance at which the photo was taken, they nevertheless suggest, in the context of exhibition communication, that they are typical museum "object identifiers" that contain further information about the object and thus present it precisely as an exhibit that can be viewed and understood (with the help of further information).[36,37] In this way, the signs provide reading cues for the museum visitors.

---

36   Heiko Hausendorf/Wolfgang Kesselheim, Die Lesbarkeit des Textes und die Benutzbarkeit der Architektur. Text- und interaktionslinguistische Überlegungen zur Raumanalyse, in: Heiko Hausendorf/Reinhold Schmitt/Wolfgang Kesselheim (eds.), Interaktionsarchitektur, Sozialtopographie und Interaktionsraum, Tübingen 2016, 55–85, see 76.

37   In this context, we believe that the table below the canvas serves as a functional equivalent to the glass showcase. Its function can be seen in ensuring that the canvas is not touched.

However, the visible arrangement of different artifacts, such as the camera, the canvas with the digital creature on it, and the keyboard included in the bar table, makes it clear to museum visitors that they are not facing a classic exhibition object that can only be looked at or a typical interactive object that can be operated.[38] Instead, the arrangement shows similarities with familiar video conference settings (e.g., via Skype), in which participants can see and interact with each other, both in a verbal and a textual manner (as the lines at the bottom of the canvas suggest). Hence, the interface design proposes the possibility of entering a mediatized interaction with the agent MAX, which already seems to be online. In this sense, MAX is presented as a potential counterpart with whom one can interact multimodally. The larger-than-life depiction of the embodied agent on the canvas and its body orientation, as well as its gaze direction, which is supposedly directed at approaching visitors, additionally create a specific form of presence that can hardly be ignored if the entity on the canvas might be a human being. That is, at first glance, the agent appears to be a present potential interlocutor who is waiting for communication partners. The camera positioned to the right of the canvas creates the expectation that the area in front of the bar table is being videotaped, so that the camera functions as the "eyes" of the digital artificial person on the canvas.

Besides the fact that MAX is an embodied agent, not a human being, the main difference compared to a "normal" video conference seems to be that the private computer screen is replaced by a large canvas (and a large camera), which are visible from afar and located in a public setting. The artifact arrangement thus indicates that possible interactions with the agent are not private but become publicly visible and presumably audible. Persons who consider interacting with MAX must therefore expect "bystanders."[39]

What do these considerations mean for the question of what actions to expect from museum visitors who approach the exhibit? Firstly, there seem to be two fundamentally different options. On the one hand, MAX—like other artifacts in the museum—is presented as an exhibit, as the signs on the tables suggest. Against this backdrop, an expectable behavior could be merely looking at the exhibit and, if necessary, getting closer to read the signs and inspect the exhibit in more detail before moving on to the next exhibit.[40] On the other hand, however, it is also possible to interpret the setting and the presence of the agent on the canvas as an invitation to

---

Its materiality ensures that no one can step into the area under the canvas, thus protecting it from contact.

38   Christian Heath/Dirk vom Lehn, Configuring 'Interactivity'. Enhancing Engagement in Science Centres and Museums, in: *Social Studies of Science* 38 (1/2008), 63–91.

39   Erving Goffman, *Forms of Talk*, Philadelphia 1981, see 132.

40   Christian Heath/Dirk vom Lehn, Configuring Reception: (Dis-)Regarding the 'Spectator' in Museums and Galleries, in: *Theory, Culture & Society* 21 (6/2004), 43–65.

engage in an interaction and treat the artificial person as a potential counterpart. In this case, the obvious action would be to go to the table with the keyboard and thus also into the camera's field of view in order to become visible to the agent system and then start engaging in a conversation either verbally or textually.

In this case, the visitors can use the signs on the tables as resources for finding possible conversation topics since the signs provide them with initial information about the agent, its capabilities, and its main developer. On the sign on the bar table, the agent is introduced as an "artificial virtual human" and the "first virtual museum guide," who can be asked questions and who gives "information about all topics" (Fig. 2; translated from German). The additional sign on the table under the screen, on the other hand, is an advertisement for a book about natural and artificial intelligence (AI) authored by MAX's "father," Professor Ipke Wachsmuth[41] (Fig. 3).

Fig. 2: Information sign, which informs about MAX and how to en- gage in interaction with the agent   *Fig. 3: Reference to a book written by the developer of MAX*

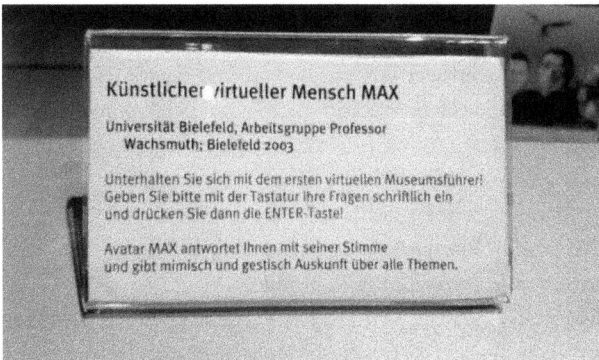

If these signs, which indicate their readability as "object identifiers," are recognized by museum visitors, they can be expected to serve as important participation cues, since they provide a possible communicative framework for interactions with MAX. This is due to the fact that the text of the sign on the bar table explicitly invites readers to pose questions to the "museum guide" MAX. If one takes this invitation and the information on the advertising sign seriously, it would make sense to formulate questions related to the museum and its exhibits or to address the topic of AI or rather MAX itself or its developer.

---

41   Ipke Wachsmuth, „Ich, Max" – Kommunikation mit Künstlicher Intelligenz, in: Tilmann Sutter, Alexander Mehler (eds.), *Medienwandel als Wandel von Interaktionsformen*, Wiesbaden 2010, 135–157.

Additionally, the small sign on the bar table not only hints at conversational topics and thus provides a frame for possible interactions, it also points out a technical feature of the interface that is not intuitively apparent from its design. Contrary to what might be expected intuitively (see above), it appears that it is not possible to communicate verbally with the agent or to switch between oral and written communication. Instead, users have to enter their utterances using the keyboard, while MAX, on the other side, provides verbal information only. This means—though it is not made explicit—that inputs in the white field can only originate with human users. Whereas in video conferences, both sides can usually write in a chat, this is not the case here, contrary to the expectations formulated above using the example of the video conference. In this sense, the sign on the table provides different reading cues as compared to the artifact arrangement that reads as a video conference setting and thus raises different visitor expectations with regard to the possibilities of interacting multimodally with MAX.

Depending on whether museum visitors have recognized the additional information on the signs and have also understood its relevance regarding the technical conditions of the exchange, different consequences for the possibilities of starting an "interaction" with MAX can be expected. If the information has been acknowledged, it is to be expected that visitors will start typing on the keyboard to formulate a question for MAX, which he can answer as a museum guide. However, if they have not taken note of and understood the information, difficulties in making contacting with the agent seem inevitable. Not only does it remain unclear what one can talk to the agent about, it appears expectable that users will attempt to make verbal contact rather than using the keyboard. Additionally, the text already visible on the upper line of the text field can be interpreted counterfactually albeit plausibly from the user's perspective as a contribution that MAX produced.

This is exactly the case in the encounter that we will analyze in the following pages. The example shows that—and how—the observed contradictory usability cues provided by the visible interface design, on the one hand, and the agent system's actual operability, on the other hand, systematically create problems that make it difficult to even begin an interaction with MAX.[42]

---

42   At least implicitly, the example also reveals another problem with the interface design (see Section 4.2). Contrary to the expected function of the camera located next to the screen, it does not help MAX perceive his environment. Instead, the camera fulfills the function of taking photos upon museum visitors' requests. During the ongoing interaction, MAX is "blind" and only registers written inputs, which he processes as an indicator of a counterpart's presence.

3.2   Interaction Analysis

The initial situation in this empirical case is shown in Figure 4. When the museum visitor approaches the exhibit, the words "Wie gehts dir? (How are you?)" are already visible on the upper line of the text field.

Fig. 4: still image of a video recording that shows the inititial situation on the canvas.

In fact, the words in the text field are the last utterance that was previously typed and sent to the machine by another person who left the site.[43] Without the necessary contextual knowledge about the interface's actual functioning, however, these words can be interpreted quite plausibly as a written contribution from the agent aimed at opening a conversation. This is exactly what happened in the present case, as the following transcript extract shows. We have given the visitor the name Didi. At the moment the transcript[44] begins, he moves towards the exit from the exhibition area

---

43   Presumably the phrase "How are you?" as the last utterance in a terminated exchange indicates that the previous interaction did not get beyond the greeting sequence before it was aborted. Our investigations at the HNF have revealed that a considerable portion of encounters with MAX—compared to "normal cases" of everyday interactions between humans—are not properly closed.

44   The simultaneously produced activities are noted on separate lines, one below the other. The sequential nature of the multimodal event becomes clear when the tables are read from left to right. Didi's verbal utterances (Didi_verb), gaze directions (Didi_gaze), and movements (Didi_move) are documented, as well as his inputs on the keyboard (Didi_type). There is also a line that documents the text that is visible on the first line

where MAX is presented and recognizes the agent that is located on the left side of the exit.

Table 1: multimodal transcript of the video (sequences 1–8)

	1	2	3	4	5	6	7	8
Didi_verb				hallo MAX. (hello MAX)		(2,7)		
Didi_gaze	*MAX*		*bar table with keyboard*		*MAX*		*bar table*	*MAX*
Didi_move			*turns left and walks towards MAX*					
screen_Text	wie gehts dir (how are you)							

As the transcript reveals, Didi first catches sight of MAX and the bar table in front of him (Columns 1 and 2) and then moves towards it. While still in motion, Didi offers a verbal greeting (Column 4) with the words, "Hello MAX". In this way, he performs the first part of a typical opening sequence of an interaction.[45,46] By talking to MAX, Didi also expresses his assumption that MAX can hear and understand him. Against this background, the final look at MAX (Column 8) can be interpreted as a signal of a "transition relevant place" that allows his interlocutor to start speaking.[47] This also indicates that a reaction from the agent can now be expected. However, as the following excerpt shows, this does not happen.

---

of the text field on the canvas (screen_text), as well as a line that notes what MAX says (MAX_verb). The exchange took place in German. We have provided English translations in brackets below the German words. For the sake of clarity, only those rows relevant to the analysis are listed in the transcript extracts.

45   Interestingly, Didi knows the agent's name. However, in the course of the encounter, it becomes obvious that he has not interacted with the system before. Presumably, he participated in a guided tour and hence previously heard the agent's name.

46   Adam Kendon, A description of some human greetings, in: *Conducting interaction. Patterns of behavior in focused encounters*, Cambridge 1990, 153–207.

47   Harvey Sacks/Emanuel A. Schegloff/Gail Jefferson, A Simplest Systematics for the Organization of Turn-Taking for Conversation, in: *Language* 50 (4/1974), 696–735.

Table 2: multimodal transcript of the video (sequences 9–14)

	9	10	11	12	13	14
Didi_verb				GU:T gehts. (I'm fine)		
Didi_gaze	MAX	text field		MAX	text field	to the left (at another exhibit)
Didi_move		stands one step in front of the bar table with the keyboard				
screen_Text	wie gehts dir (how are you)					

There is no visible reaction from the agent. Didi, who has meanwhile come to a stop one step in front of the bar table (Column 9), directs his gaze to the text field (Column 10) and then utters the phrase "GU:T geht's (I'm fine)" (12), which can be clearly interpreted as an answer to the question "How are you?" that remains in the text field. Obviously, Didi assumes that the text in the field is a contribution from MAX and possibly also a meaningful reaction to his previous greeting. Counterfactually but quite expectedly (as stated above in Section 3.1), Didi's response expresses the assumption that MAX communicates in written language and can also hear. This is confirmed by Didi's eye movement, which initially moves away from the text field and towards the embodied agent (Column 11), which he addresses verbally, only to turn back to the text field while speaking (Column 13), where an answer is apparently expected.

However, the text box remains unchanged, so Didi begins to look at the artifact's wider surroundings (Column 14). This can be interpreted as a search for clues that he can use as resources to understand what is going wrong and why MAX is not responding. As the following transcript excerpt shows, this attempt is successful.

Table 3: multimodal transcript of the video (sequences 15–18)

	15	16	17	18
Didi_gaze	MAX	sign on the bar table	keyboard	
Didi_move				steps up to the bar table and moves his hands towards the keyboard

While searching for reading cues that might help him to understand what has gone wrong, Didi's gaze wanders to the sign (Column 16) and then to the keyboard (Column 17) on the bar table, which causes him to move even closer to the table and take his hands out of his pockets (Column 18). In this way, Didi prepares for a change of communication mode, that is, switching from verbal to written communication. After taking his hands out of his pockets, he puts them on the keyboard and starts typing (Column 19).

Table 4: multimodal transcript of the video (sequences 19–23)

	19	20	21	22	23
Didi_gaze		*text field*		*keyboard*	*text field*
Didi_type	g g u t (g g o o d)	(3.3)			

At this point, not only can we observe a change of communication mode but also a change of orientation: Whereas Didi previously oriented himself towards a common perceptual space with MAX (i.e., turning to the interface, looking at it, addressing the agent), he is now oriented towards (at least, on his part) purely text-based communication. His gaze switches back and forth between the text field and the keyboard for about half a minute, while ignoring the embodied agent on the canvas. After he reads the additional information about how the communication interface technically functions, he adapts his behavior to the now recognized technical restrictions of the "interaction" with MAX and thus proves his "architectural literacy."[48]

First, Didi types "g g u t (g g o o d)" (Column 19), which can be interpreted as an (erroneous) elliptical form repeating what has been said before. That is, Didi corrects his previous behavior in a technical sense by translating it into a communication mode perceptible to MAX. However, he does not send the answer but first looks at the written letters (Column 20) and then to the keyboard and back to the text field (Columns 22 and 23). With these eye movements, Didi stops his current writing activities, and the question of what will happen next arises. It is conceivable that he recognizes the spelling error when looking at the text field (Column 20) and starts a corresponding self-repair in the next step.

---

48   Hausendorf/Schmitt, Architecture-for-interaction, 442.

Table 5: multimodal transcript of the video (sequences 24–30)

	24	25	26	27	28	29	30
Didi_gaze	text field		keyboard	text field			keyboard
Didi_type	<<deletes> t u g>	(4.0)			<<deletes> g>	(2.5)	

The expected behavior seems to be realized. Didi deletes the three letters t, u, and g (Column 24), leaving only the letter g in the text field. This suggests that he is planning a spelling correction and plans to enter the word "gut (good)" correctly again. However, after a long pause of four seconds (Column 25), Didi also deletes the remaining letter (Column 28), whereupon another pause occurs (Column 29). It can be assumed that he will now start again by not only making a spelling correction but also choosing different words.

Table 6: multimodal transcript of the video (sequences 31–35)

	31	32	33	34	35
Didi_gaze		text field		keyboard	
Didi_type	s e h r (v e r y)	(3,8)			s c h l e c h t <Enter> (b a d)

As the transcript shows, the expected behavior actually happened. Didi enters the words "Very bad" (Columns 31 and 35), with a pause of almost four seconds between entering the first and second word (Column 33). During the pause, he directs his gaze first to the text field and then back to the keyboard (Columns 32 and 34). It seems as if he is considering how to continue his turn. This indicates a certain indecisiveness and possibly also uncertainty as to what to enter. At the same time, Didi's hesitation and the long pauses between his typing activities indicate that he no longer feels any practical pressure to act immediately, as would be the case in face-to-face situations, in which the interlocutors mutually recognize each other.[49] Whereas Didi spontaneously answered in the verbal communication mode with the very common phrase ("GU:T geht's (I'm fine)" (Column 12)), he now makes use of a behavior that is typical in written online communication such as chatting

---

49   Harold Garfinkel, *Studies in ethnomethodology*, Englewood Cliffs 1967, see 12.

or instant messaging, namely "message construction repair."[50] However, whereas in most cases, people "edit their posts when responding to something posted by their co-participant,"[51] the situation here is different. Didi does not adapt to his counterpart's new activities but rather decides to break with the normative expectations that are commonly applied in greeting sequences.[52]

An expression of bad health, especially using the strong variant Very bad," is unusual and would only be expected in interactions between people who are familiar with each other and hence seriously share their individual mood with one another. In view of this, Didi's corrected answer can only plausibly be interpreted as an explicit distancing from the common normative expectations that exist in everyday conversations. In doing so, he indicates to potential bystanders in the background that he is not seriously engaged in an exchange with the artificial counterpart. Additionally, this breach of everyday expectations can also be interpreted as starting a test of the artifact's communicative abilities. The questions now arise as to whether the embodied agent will recognize the violation of basic communication norms and how it will react.

Table 7: multimodal transcript of the video (sequences 36–37)

	36	37
MAX_verb	(3.0)	ich spüre negative schwingungen
		(I feel negative vibes)

Again, it takes a moment for something to happen (Column 36). After three seconds, however, MAX responds with the words "I feel negative vibes" (Column 37). In this, he follows Didi's input in terms of content but does so in a highly specific way. Whereas from Didi's perspective, the words "Very bad" refer unseriously to his general condition, MAX's statement refers to an atmospheric disturbance in the relationship between the two interlocutors. This could be interpreted as a humorous reaction to Didi's contribution, with which MAX indicates that he has understood the lack of seriousness and thus proves his communicative competence. However, it is also possible to interpret Max's response as inappropriate and accordingly as an indication of limitations in MAX's communicative abilities.

---

50   Joanne Meredith/Elizabeth Stokoe, Repair, Comparing Facebook 'chat' with spoken interaction, in: *Discourse & Communication* 8 (2/2014), 181–207.

51   Meredith/Stokoe, Repair: Comparing Facebook 'chat' with spoken interaction, 250.

52   Jack Sidnell, *Conversation analysis. An introduction*, Chichester 2010, see 208.

Human users can only speculate which of the interpretations is appropriate and react accordingly on a test basis. Against this, in our analysis, we have the option of opening the "black box" of MAX's operability to take a look at the agent's internal processes. We will now do this with the first utterance the agent produced during the encounter under investigation. This will allow us to reveal how MAX's internal processes operate and elucidate the consequences in the course of the interaction.

4. Analysis of MAX's "Internal" Architecture-For-Interaction

In addition to the analysis of MAX's interface, the second part of the analysis is dedicated to the agent's internal processes to show how the technical system's programming opens up and limits the interaction possibilities. For this, we examine programming scripts that provide information about the agent's internal system architecture and its operability. The analysis shows how user inputs are classified and examined based on the agent's system architecture and thus provides insights into how MAX "thinks" and acts based on its programming.

4.1   MAX's Internal System Architecture

Essentially, MAX's system architecture follows the so-called belief–desire–intention (BDI) concept.[53] That is, the agent has beliefs about the world (and his counterpart) and also has desires and intentions. A so-called "BDI interpreter" determines the agent's behavior. This interpreter is basically responsible for pursuing

> multiple plans (intentions) to achieve goals (desires) in the context of up-to-date world knowledge (beliefs). [...] Most of the plans implement condition-action rules [...]. Such rules can test either the user input (text, semantic or pragmatic aspects) or the content of dynamic knowledge bases (beliefs, discourse or user model); their actions can alter the dynamic knowledge structures, raise internal goals and thus invoke corresponding plans, or trigger the generation of an utterance.[54]

Regarding the agent's plans, a distinction must be made between global (top-level) meta-plans and plans that relate to local (low-level) dialogue goals. The agent can pursue the global goal of finding out his counterpart's name while simultaneously determining, through concrete dialogue (low level), that the user is not currently interested in giving their name. This enables the agent to put its overarching dialogue

53   Marcus J. Huber, JAM: A BDI-theoretic mobile agent architecture, in: *Proceedings of the third annual conference on Autonomous Agents*, Seattle, Washington 1999, 236–243.

54   Kopp et al., A Conversational Agent as Museum Guide, 333–34.

goal on hold to wait for a suitable moment to implement the meta-plan.[55] Technically, the hierarchization of goals is realized by a utility function that rates MAX's several desires, and the highest rated one is selected at a certain moment to become the current goal.

In addition to this "cognitive" architecture, MAX has an emotion component that determines the agent's current mood, which also influences the formation of intentions. In this way, the agent can react, for instance, to disappointed expectations resulting from the fact that certain (initially cognitively raised) dialogue goals (such as getting the interlocutor's name) have not been achieved. These mood changes also become visible to MAX's counterparts, for they are indicated by certain facial expressions and gestures made by the embodied agent.

It is also important for MAX to possess both static and dynamic knowledge. That is, the system possesses world knowledge that has been programmed into it. Additionally, it generates and stores dynamic knowledge gleaned directly from interaction processes, which—at least, if everything works—is cached and deleted again with every new conversation. The dynamic knowledge components "user models"[56] and "discourse models" are important for how the agent adapts its activities to its counterpart and to the interaction's previous course.

In concrete terms, the system architecture works such that, on the one hand, the agent proactively generates statements according to its own plans, and on the other hand, interlocutors' utterances are registered and stored and used to change the agent's internal state and respond to the preceding turn. For this purpose, utterances are analyzed for their communicative functions based on a certain preprogrammed scheme. According to the system architecture, communicative functions always consist of three components: a performative component, a reference level, and content. Regarding the performative aspect, the system examines whether an utterance provides or requests information. Regarding the reference level, the system distinguishes between three different levels of dialogue: the interaction level, the discourse level, and the content level. Whereas actions on the interaction level, are concerned with the opening, continuation, and closing of the interaction, actions on the discourse level deal with interaction management; that is, they manage "the topic and flow of conversation (e.g., the suggestion of a new topic to talk about). At the content level, information about the current topic is conveyed."[57] Regarding

---

55   Kopp et al., A Conversational Agent as Museum Guide, 336–37.

56   MAX's counterparts are represented in list form in the user model. That is, they are considered as persons with certain characteristics (e.g., in the dimensions of age, name, gender, profession, etc.). During interactions, the agent tries to populate the list and thus mobilizes users as resources for filling empty "slots" within its user model.

57   Kopp et al., A Conversational Agent as Museum Guide, 334.

the third component (content), the question concerns what concrete content a particular utterance provides or what exact information is being requested.

Communicative functions are stored in the system in the form <performative>.<reference level>.<content>[arguments] and, if necessary, are supplemented with additional arguments in parentheses. For example, the system will interpret an utterance such as "Hello" as "provide.interaction.greeting," whereas a statement such as "Let's talk about football" is processed as "askFor.discourse.topic.sports."[58] Depending on the system's current state, every input leads to the application of the most appropriate plan for the situation, which is then executed, generating a corresponding output.

How exactly the interpretation of utterances and the execution of plans work on the agent's side is the subject of the following considerations, as we proceed with the interaction analysis. We start with the first utterance recognizable to the agent, that is, "Very bad," and show how the system interprets it.

4.2   Continuation of Interaction Analysis

Table 8: transcript of the internal processes of MAX (line 1)

Didi_text
[1]sehr schlecht (very bad)

By sending the input "Very bad" (Table 8, Line 1) to the system, Didi (from his perspective) has responded unseriously to a question ("How are you?"), which, in fact, the agent never asked (see Section 3.2). We know that the agent responded to Didi's input with the utterance "I feel negative vibes," but up till now, we did not understand how the agent system generated this statement. The system's internal processes provide information on this.

---

58   Kopp et al., A Conversational Agent as Museum Guide, 335.

Table 9: transcript of the internal processes of MAX (line 2)

Max_BDI[59]
[2]po129: 7.333333333333334 – curInput::!askFor(disliking) – smalltalk.disliking

As a look into the machine's internal processes reveals, it assigns the communicative function "askFor(disliking).smalltalk.disliking" to the current input ("curInput") (Table 9, Line 2). That is, the preceding statement "Very bad" is not interpreted as a response that provides information about its author's wellbeing during a greeting sequence; instead, the utterance is interpreted as a request (combined with the optional argument "disliking") in the context of ongoing small talk. This means that according to Max's interpretation, Didi's utterance expresses a negative positioning regarding the actual content of the conversation.[60] Given the observable situation in front of the canvas, it appears that MAX's interpretation seems to be inappropriate. As the next extract shows (Table 10), this has significant consequences for the further course of the interaction.

---

59   Here, we introduce a new line in our transcription scheme. Max_BDI shows program scripts to provide insights into the system's internal processes.

60   Small talk is located at the level of the content of the "interactions" (Kopp et al., A Conversational Agent as Museum Guide, 337).

Table 10: transcript of the internal processes of MAX (lines 3–19)

Max_BDI

```
[3] <rule name="smalltalk.disliking" utility="-2">
[4] <match>
[5] <convfunction type="!askFor"
[6] modifier="disliking"/>
[7] </match>
[8] <action>
[9] <command function="trigger-emotions"
[10] arguments="SPONTANEOUS -20"/>
[11] <random>
[12] <act function="provide.discourse.disagree">Ich
[13] spuere negative Schwingungen.</act>
[14]<act>Lass uns lieber ueber was sprechen was du
[15]magst.
[16] </act>
[17] </random>
[18] </action>
[19] </rule>
```

The extract shows that the agent selects a plan of action that seems appropriate to this interpretation of the situation—but it is not appropriate to the situation in which Didi finds himself. The actualized and executed plan is called "smalltalk.disliking," and it defines both certain changes in the system's internal state and an output with a certain communicative function.

First, the action plan's utility value decreases by two points when it is activated (Table 10, Line 3). This changes MAX's internal priorities and action goals. Additionally, the command function "trigger emotions" (Table 10, Lines 9–10) is used to define a certain facial expression that the embodied agent performs when the plan is executed.[61] Further, the plan determines MAX's reaction. The system produces an utterance that fulfills the function "provide.discourse.disagree" (Table 10, Line 13), which is randomly selected from two possible formulations (Table 10, Lines 11–17) and then executed.

---

61 Christian Becker/Stefan Kopp/Ipke Wachsmuth, Simulating the Emotion Dynamics of a Multimodal Conversational Agent, in: Takeo Kanade/Josef Kittler/Jon M. Kleinberg/ Friedemann Mattern/John C. Mitchell/Oscar Nierstrasz et al. (eds.), *Affective Dialogue Systems* (vol. 3068), Berlin, Heidelberg 2004, 154–165.

Table 11: transcript of the internal processes of MAX (lines 20–21)

Max_text

       [20]Ich spuere negative Schwingungen. (I feel negative vibes)

Max_BDI

       [21]<act function="provide.discourse.disagree" emphasis="none">Ich spuere negative Schwingungen.</act>

In the present case, the answer is "I feel negative vibes" (Table 11, Line 20). As described, this answer to Didi's input makes only limited sense. The agent reacts as if there were factual differences at the level of small talk, although Didi apparently gave an unserious answer to the question "How are you?" Accordingly, the extract shows precisely how the agent's "inner world" and the communicative "outside world" decouple from each other, which results in communicative problems that become evident in the further course of the encounter. First, we will take a look at MAX's processes after the system has responded to Didi's previous utterance (Table 12).

Table 12: transcript of the internal processes of MAX (lines 22–59)

Max_bdi

```
[22]go016: get-name – hold-initiative – goals.user.getName

[23]<rule name="goals.user.getName" utility="10">
[24]<goal name="get-name" context="emptyslot name" />
[25]<action function="take-initiative">
[26][...]
[27]</action>
[28]<action function="hold-initiative">
[29]<switch var="$cycles">
[30]<cond value="1">
[31][...]
[32]</cond>
[33]<cond value="2,3">
[34]<random>
[35]<act>Jetzt aber weiter mit deinem Namen.</act>
[36]</random>
[37]<random>
[38]<block>
[39]<act>Du kannst Dir ja auch einen Namen ausdenken wenn Du
[40]Deinen nicht sagen willst.</act>
[41]<act function="askFor.content.name">Also?</act>
[42]</block>
[43]<act function="askFor.content.confirmation">Willst Du ihn
[44]wirklich nicht sagen? Ich verrat es auch nicht weiter.
[45]Ok?</act>
[46]<act function="askFor.content.name">Wie hat Dich Deine Mutter frueher immer
      genannt?</act>
[47]<act function="askFor.content.confirmation">Hast Du
[48]vielleicht einen Spitznamen?</act>
[49]</random>
[50]</cond>
[51]<else>
[52][...]
[53]</else>
[54]</switch>
[55]</action>
[56]<action function="resume-initiative">
[57][...]
[58]</action>
[59]</rule>
```

The extract shows MAX (re-)activating its global dialogue goal of obtaining the interlocutor's name (Table 12, Line 22), which had been put on hold. The "hold-initiative" information signals that MAX maintained this goal, which means that the agent wanted to achieve it before but put it on hold to respond to user input in the meantime. The fact that the goal was put on hold indicates that from the system's perspective, Didi is not a new user but rather the same person that entered the question "How are you?" Obviously, MAX has not yet deleted its dynamic knowledge regarding the last interaction and hence "believes" that it is still engaged in that particular interaction. This explains why the agent did not previously interpret Didi's opening statement as part of an (unserious) greeting act but rather as a substantive contribution to ongoing small talk.[62] Consequently, MAX appears to be "in the wrong movie," for the agent system has mistakenly assumed that it already made several unsuccessful attempts to find out its counterpart's name (Table 12, Line 33), which leads the agent to start a new attempt.

Table 13: transcript of the internal processes of MAX (lines 60–61)

Max_text
[60]Jetzt aber weiter mit deinem Namen. (lets continue with your name.) [61]Hast Du vielleicht einen Spitznamen? (do you have a nickname?)

According to its current dialogue goal, the agent produces an output comprising two parts (Table 13). The first part is "Let's continue with your name" (Table 12, Line 35; Table 13, Line 60), and the second part is randomly selected from among various possible utterances (Table 12, Lines 39–49). In the present case, the output is "Do you have a nickname?" (Table 12, Line 47f; Table 13, Line 61), which, according to the system architecture, fulfils the communicative function "askFor.content.confirmation."[63]

MAX's interpretation of the situation still differs significantly from Didi's actual situation in front of the canvas. However, given the fact that Didi's last utterance was a response to the question "How are you?" and thus appears as a(n) (unserious) reaction within a greeting sequence, MAX's utterances could be interpreted counterfactually (given MAX's internal processes) as a rude first attempt to find out Didi's

---

62   This also indicates that the system does not use the camera to register when people enter and leave the interaction space in front of the bar table.

63   As can be seen below, this is a bit odd because according to this function, a confirmation is expected, not the provision of a name (Fig. 20, Lines 64–80).

name during an opening sequence that appears to be a bit strange. That is, from Didi's perspective, MAX's utterance might somehow fit the situation, although the agent is, in fact, misguided.

The question now concerns how Didi responds to MAX's statement. Will he provide his name, or will he proceed in a provocative manner?

Table 14: transcript of the internal processes of MAX (line 62)

Didi_text
[62]Didi

As the snippet shows (Table 14), Didi overlooks the rudeness expressed during MAX's turn and enters "Didi" (Line 62). In doing so, he indirectly answers the question of *whether* he has a nickname by entering this nickname. Thus, he does not simply answer technically with "yes" or "no," he also substantially responds to the first part of the request. In doing so, from his perspective, Didi fulfils the expectations associated with the question of whether he has a nickname. However, although this kind of response would be unproblematic and adequate in normal interactions between humans, it poses problems for MAX, since the agent expects a clear "yes" or "no" answer and cannot process the implicit "yes" hidden in Didi's utterance. The next snippet makes this obvious (Table 15).

Table 15: transcript of the internal processes of MAX (line 63)

max_bdi
[63]po106: smalltalk.fallback.repeatAskFor

According to MAX's interpretation, no confirmation (positive or negative) has been given in response to the preceding question. Consequently, a "fallback rule" is triggered (Table 15, Line 64), which aims to repeat the preceding request to get a definitive "yes" or "no" (Table 16, Lines 70–77).

Table 16: transcript of the internal processes of MAX (lines 64–80)

Max_bdi

```
[64]<rule name="smalltalk.fallback.repeatAskFor" utility="-
15">
[65][...]
[66]<action>
[67]<command function="add-context" arguments="repeated-yn-
[68]question ultrashort"/>
[69]<random>
[70]<act function="$lastFunc">Was soll das denn heissen – Ja
[71]oder Nein?</act>
[72]<act function="$lastFunc">Heisst das jetzt Ja oder
[73]Nein?</act>
[74]<act function="$lastFunc">Ist dass ein Ja oder ein
[75]Nein?</act>
[76]<act function="$lastFunc">Was willst du damit sagen – Ja
[77]oder Nein?</act>
[78]</random>
[79]</action>
[80]</rule>
```

Parallel to the actual goal of obtaining confirmation, the global meta-goal—namely, to find out the user's name—still remains upheld, as the next snippet reveals (Table 17, Line 81).

Table 17: transcript of the internal processes of MAX (line 81)

Max_bdi

```
[81]g0016: get-name – hold-initiative – goals.user.getName
```

Consequently, MAX executes its short- and long-term plans in succession and hence produces output comprising utterances generated by the "smalltalk.fall-back.repeatAskFor" (Table 16, Lines 70–77) rule *and* the "user.getName" (Table 12, Lines 35–48) rule (Table 18, Line 82).

Table 18: transcript of the internal processes of MAX (line 82)

Max_text
[82]Was willst du damit sagen – Ja oder Nein? Jetzt aber weiter mit deinem Namen. Willst Du ihn wirklich nicht sagen? Ich verrat es auch nicht weiter. Ok? (What are you trying to say – yes or no? Let's continue with your name. Do you really not want to say it? I won't tell anyone. OK?)

With this output, it becomes explicit that MAX has treated Didi's previous utterance, in which he provided his nickname, as his refusal to answer. Although this makes sense in the agent's operational logic, the response must come unexpectedly for Didi. Against this backdrop, the question concerns how Didi will react to this surprising and perhaps also unsatisfying and confusing reaction. On the one hand, it seems possible that he will adapt to the agent's limited communication capabilities, which are increasingly becoming apparent. In this case, he could perform a self-repair and answer more explicitly. For example, he could say "Sorry. My nickname is Didi." Another option is to express his dissatisfaction with the system's performance in one way or the other because its capabilities do not live up to his expectations.

Table 19: transcript of the internal processes of MAX (line 83)

Didi_move
[83]leaves the bar table

As the last extract from the transcript reveals (Table 19), Didi chooses the second option. Instead of responding once more, he breaks off the interaction and leaves the bar table. In doing so, he shows that he is no longer interested in continuing the conversation and hence expresses his dissatisfaction. Additionally, this behavior conveys that he does not view MAX as a serious counterpart, for in a "normal" interaction, it would be very rude to leave without any kind of closing and/or farewell.[64] In

---

64   Regarding the analysis of the system architecture, MAX noteworthily continues to ask for its counterpart's name for some time, until the agent finally abandons that task and utters a farewell phrase—a long time after Didi has left the scene. This again indicates that the agent has no capabilities to recognize what is happening in its environment except for analyzing input sent to the system via the keyboard.

sum, the analysis reveals that not only does the interface design evoke communicative problems, but the agent's system architecture systematically produces trouble due to its limitations. Strikingly, we cannot identify any time when there exists the possibility of establishing a shared situation definition and thus "common ground"[65] between the machine and the user. Unsurprisingly, this results in unresolvable communicative problems.

5. Conclusion

The systematical intertwining of the analysis of the agent system's architecture-for-interaction with the empirical analysis of a factual human–machine encounter reveals how the various elements of the system's architecture-for-interaction each individually but also in combination contribute to opening up and, in the presented case, to, above all, restricting and complicating the interaction possibilities. This begins with the fact that MAX's interface seems, at first glance, to allow verbal interaction on both sides, which, in fact, is not the case and has yet to be figured out by the system's human users. The difficulties continue with the problem that stored text in the text field can mistakenly be interpreted as output from MAX, not as input from previous users, which happened in the presented case. Additionally, the agent's limited perceptual capabilities lead to internal interpretations of the situation that do not correspond with the actual situation unfolding in front of the canvas. All these issues create a cascade of persistent problems that cannot be fixed in the course of the encounter.

If one generalizes these insights, the empirical results first confirm that architectures-for-interaction systematically pre-structure the possibilities for interaction with communicative AI systems. Accordingly, the systematic consideration of technical architectures-for-interaction in empirical analyses systematically contributes to identifying the limits, problems, and possibilities of human–machine communication. Of particular relevance is that the consideration of architectures-for-interaction clearly reveals where exactly and for what reasons communicative problems occur. Thus, the analysis of architectures-for-interaction can contribute not only to an adequate understanding of human–machine communication but also to the evaluation of communicative AI systems and their optimization. A necessary condition for this is, of course, the possibility of gaining access to corresponding systems' programming, which is unlikely, especially for commercial systems. In this

---

65   Keith Allan, What is Common Ground?, in: Alessandro Capone/Marco Carapezza/Franco Lo Piparo (eds.), *Perspectives on linguistic pragmatics* (vol. 2), Cham, New York 2013, 285–310.

sense, more "transparency in artificial intelligence"[66] would also be desirable with regard to communicative AI systems' architectures-for-interaction.

Bibliography

Allan, Keith, What is Common Ground?, in: Alessandro Capone/Marco Carapezza/ Franco Lo Piparo (eds.), *Perspectives on linguistic pragmatics (vol. 2)*, Cham, New York 2013, 285–310.

Arminen, Ilkka/Licoppe, Christian/Spagnolli, Anna, Respecifying Mediated Interaction, in: *Research on Language and Social Interaction* 49 (4/2016), 290–309.

Bearne, Eve/Kress, Gunther, Editorial, in: *Reading: literacy and language* 35 (3/2001), 89–93.

Becker, Christian/Kopp, Stefan/Wachsmuth, Ipke, Simulating the Emotion Dynamics of a Multimodal Conversational Agent, in: Takeo Kanade/Josef Kittler/Jon M. Kleinberg/Friedemann Mattern/John C. Mitchell/Oscar Nierstrasz et al. (eds.), *Affective Dialogue Systems (vol. 3068)*, Berlin, Heidelberg 2004, 154–165.

Belpaeme, Tony/Kennedy, James/Ramachandran, Aditi/Scassellati, Brian/Tanaka, Fumihide, Social robots for education: A review, in: *Science robotics* 3 (21/2018).

Bickmore, Timothy/Pfeifer, Laura/Schulman, Daniel, *Relational Agents Improve Engagement and Learning in Science Museum Visitors*, Boston 2011.

Garfinkel, Harold, *Studies in ethnomethodology*, Englewood Cliffs 1967.

Gaver, William W., Technology affordances, in: Scott P. Robertson (ed.), *Reaching through technology. CHI '91; conference proceedings; New Orleans Louisiana April 27– Mai 2 1991. the SIGCHI conference*, New Orleans 1991, 79–84.

Ghafurian, Moojan/Hoey, Jesse/ Dautenhahn, Kerstin, Social Robots for the Care of Persons with Dementia, in: *ACM Trans. Hum.-Robot Interact.* 10 (4/2021), 1–31.

Gibson, James J., The Theory of Affordances, in: Robert Shaw/John Bransford (eds.), *Perceiving, acting, and knowing. Toward an ecological psychology*, Hillsdale 1977, 67–82.

Goffman, Erving, *Forms of Talk*, Philadelphia 1981.

Guzman, Andrea L./Lewis, Seth C., Artificial Intelligence and Communication: A Human–Machine Communication Research Agenda, in: *New Media & Society* 22 (1/2020), 70–86.

Hausendorf, Heiko/Kesselheim, Wolfgang, Die Lesbarkeit des Textes und die Benutzbarkeit der Architektur. Text- und interaktionslinguistische Überlegungen zur Raumanalyse, in: Heiko Hausendorf/Reinhold Schmitt/Wolfgang Kessel-

---

66   Stefan Larsson/ Fredrik Heintz, Transparency in artificial intelligence, in: *Internet Policy Review* 9 (2/2020).

heim (eds.), *Interaktionsarchitektur, Sozialtopographie und Interaktionsraum*, Tübingen 2016, 55–85.

Hausendorf, Heiko/Schmitt, Reinhold, Standbildanalyse als Interaktionsanalyse: Implikationen und Perspektiven, in: Heiko Hausendorf/Reinhold Schmitt/Wolfgang Kesselheim (eds.), *Interaktionsarchitektur, Sozialtopographie und Interaktionsraum*, Tübingen 2016, 161–187.

Hausendorf, Heiko/Schmitt, Reinhold, Architecture-for-interaction: Built, designed and furnished space for communicative purposes, in: Andreas H. Jucker/Heiko Hausendorf (eds.), *Pragmatics of Space*, Berlin, Boston 2022, 431–472.

Hausendorf, Heiko/Schmitt, Reinhold/Kesselheim, Wolfgang (eds.), *Interaktionsarchitektur, Sozialtopographie und Interaktionsraum*, Tübingen 2016.

Heath, Christian/vom Lehn, Dirk, Configuring Reception: (Dis-)Regarding the 'Spectator' in Museums and Galleries, in: *Theory, Culture & Society* 21 (6/2004), 43–65.

Heath, Christian/vom Lehn, Dirk, Configuring 'Interactivity'. Enhancing Engagement in Science Centres and Museums, in: *Social Studies of Science* 38 (1/2008), 63–91.

Holland, Jane/Kingston, Liz/McCarthy, Conor/Armstrong, Eddie/O'Dwyer, Peter/Merz, Fionn/McConnell, Mark, Service Robots in the Healthcare Sector, in: *Robotics* 10 (1/2021), 47.

Huber, Marcus J., JAM: A BDI-theoretic mobile agent architecture, in: *Proceedings of the third annual conference on Autonomous Agents*, Seattle, Washington 1999, 236–243.

Hutchby, Ian, Technologies, Texts and Affordances, in: *Sociology* 35 (2/2001), 441–456.

Kendon, Adam, A description of some human greetings, in: *Conducting interaction. Patterns of behavior in focused encounters*, Cambridge 1990, 153–207.

Knappett, Carl/Malafouris, Lambros (eds.), *Material agency. Towards a non-anthropocentric approach*, New York 2008.

Kopp, Stefan/Gesellensetter, Lars/Krämer, Nicole C./Wachsmuth, Ipke, A Conversational Agent as Museum Guide – Design and Evaluation of a Real-World Application, in: Themis Panayiotopoulos/Jonathan Gratch/Ruth Aylett/Daniel Ballin/Patrick Olivier/Thomas Rist (eds.), *Intelligent Virtual Agents*, Springer Berlin, Heidelberg 2005, 329–343.

Krummheuer, Antonia, Zwischen den Welten. Verstehenssicherung und Problembehandlung in künstlichen Interaktionen von menschlichen Akteuren und personifizierten virtuellen Agenten. In: Herbert Willems (ed.), *Weltweite Welten. Internet-Figurationen aus wissenssoziologischer Perspektive* (1st edition), Wiesbaden 2008, 269–294.

Larsson, Stefan/Heintz, Fredrik, Transparency in artificial intelligence, in: *Internet Policy Review* 9 (2/2020).

Latour, Bruno, *Reassembling the social. An introduction to actor network theory*, Oxford 2005.

Marino, Mark C., *Critical code studies*, Cambridge, Massachusetts 2020.

Martin, David/Bowers, John/Wastell, David, The Interactional Affordances of Technology. An Ethnography of Human-Computer Interaction in an Ambulance Control Centre, in: Harold Thimbleby/Brid O'Conaill/Peter J. Thomas (eds.), *People and Computers XII. Proceedings of HCI '97*, London 1997, 263–281.

Mayer, Henning/Muhle, Florian/Bock, Indra, Whiteboxing MAX. Zur äußeren und inneren Interaktionsarchitektur eines virtuellen Agenten, in: Eckhard Geitz/Christian Vater/Silke Zimmer-Merkle (eds.), *Black Boxes – Versiegelungskontexte und Öffnungsversuche. Interdisziplinäre Perspektiven* (1st edition), Berlin 2020.

Meredith, Joanne/Stokoe, Elizabeth, Repair, Comparing Facebook 'chat' with spoken interaction, in: *Discourse & Communication* 8 (2/2014), 181–207.

Mubin, Omar/Stevens, Catherine J./Shahid, Suleman/Mahmud, Abdullah Al/Dong, Jian-Jie, A Review of the Applicability of Robots in Education, in: *Technology for Education and Learning* 1 (1/2013).

Muhle, Florian/Bock, Indra, Intuitive Interfaces? Interface Design and its Impact on Human-Robot Interaction, in: *Mensch und Computer 2019 – Workshopband*, Bonn 2019, 346–347.

Pitsch, Karola, Referential Practices for a Museum Guide Robot. Human-Robot-Interaction as a Methodological Tool to Investigate Multimodal Interaction, in: *Mensch und Computer 2019 – Workshopband*, Bonn 2019.

Reeves, Stuart/Porcheron, Martin/Fischer, Joel, "This is Not What We Wanted": Designing for Conversation with Voice Interfaces, in: *Interactions* 26 (1/2018), 46–51.

Sacks, Harvey/Schegloff, Emanuel A./Jefferson, Gail, A Simplest Systematics for the Organization of Turn-Taking for Conversation, in: *Language* 50 (4/1974), 696–735.

Schegloff, Emanuel A./Sacks, Harvey, Opening up closings, in: *Semiotica* 8 (4/1973), 289–327.

Schrock, Andrew Richard, Communicative affordances of mobile media: Portability, availability, locatability, and multimediality, in: *International Journal of Communication* 9 (2015), 18.

Sidnell, Jack, *Conversation analysis. An introduction*, Chichester 2010.

Suchman, Lucy A., *Human-machine reconfigurations. Plans and situated actions* (2nd edition), Cambridge 2007.

Wachsmuth, Ipke, „Ich, Max" – Kommunikation mit Künstlicher Intelligenz, in: Tilmann Sutter, Alexander Mehler (eds.), *Medienwandel als Wandel von Interaktionsformen*, Wiesbaden 2010, 135–157.

Zillien, Nicole, Die (Wieder-)Entdeckung der Medien – Das Affordanzkonzept in der Mediensoziologie, in: *Sociologia Internationalis* 46 (2/2008), 161–181.

Part III: Smart Speakers in (Inter-)Action

VUI-Speak: There Is Nothing Conversational about "Conversational User Interfaces"

Brian L. Due, Louise Lüchow

Abstract *In this chapter, we suggest a concept for describing participants' practices regarding progressively adapting their actions to fit the computational system in voice user interfaces (VUIs) such as Google Home. We describe this phenomenon as "VUI-speak." Although developers aim at enabling computers to communicate like humans, our study shows that, on the contrary, people accommodate the device through VUI-speak. Based on video ethnographic studies and ethnomethodological conversation analysis (EM/CA) of blind people's natural use of Google Home, this research contributes to EM/CA studies of human–computer interaction, human–robot interaction, and VUIs in particular. The research findings suggest (1) that VUI-speak is produced at the third position in a five-part sequential structure, (2) that a change in action formation occurs, and (3) that this change relates to producing what we call an "application-oriented turn." This research has practical implications for the design of conversational systems and contributes to the expanding field of EM/CA research on VUI interaction.*

1. Introduction

Mainstream voice user interface (VUI) systems constitute a rapidly evolving field involving all the major IT companies, including Google, Apple, Microsoft, and Amazon. Sales of devices with digital assistants, such as Amazon's Alexa and Google Home, are currently doubling every two years,[1] and they have become more common in Denmark, the country on which the results reported in this chapter are based. However, although the developers' stated aim is to enable "computers to

---

[1] Statista, Google Home Global Shipments 2016–2025, Statista, 2020. URL: https://www.s tatista.com/statistics/1022722/worldwide-google-home-unit-shipment/ [last accessed: August 15, 2023].

communicate like humans,"[2,3] our research on the naturally occurring, everyday use of such devices shows that the exact opposite is the case in reality.

Whereas in ideal settings, user tests in experimental settings can produce results where humans might talk "naturally" with an AI device for a short period of time, this "naturalness" is restricted to only one particular kind of social action: question and answer sequences.[4] One of the bigger obstacles to the progress of conversational VUIs is to adhere to the actual, natural ways in which people produce verbal actions in sequences.[5] In this paper, we show that because only a limited number of actions work as commands in current technologies,[6] people attune the way they speak to suit the computational system. This paper focuses on and demonstrates the *haecceity*—the particularity, the *thisness*[7]—of how people accommodate talking to a system in situated encounters with the technology. Although comprehensive research on VUIs exists, few studies have empirically examined everyday natural use of these systems. Consequently, little is known about practical interaction with VUIs. That participants encounter problems conversing with VUIs is not a new finding. Rather, our new contribution to this field is terminology to denote the consequences these problems have for human language production, which becomes computerized into what we describe as "VUI-speak."

This work is in the tradition of human–computer interaction (HCI), building on ethnomethodology and conversation analysis (EM/CA),[8] based on video recordings

---

2   Cathy Pearl, *Designing Voice User Interfaces: Principles of Conversational Experiences* (1st edition), Beijing 2017.

3   Cf. Cynthia Breazeal/Kerstin Dautenhahn/Takayuki Kandai, Social Robotics, in: Bruno Siciliano/Oussama Khatib (eds.), *Springer Handbook of Robotics*, Berlin; Heidelberg 2016, 1935–71.

4   Martin Porcheron et al., Voice Interfaces in Everyday Life, in: *Proceedings of the 2018 CHI Conference on Human Factors in Computing Systems* 640, New York 2018, 1–12.

5   Saul Albert/William Housley/Elizabeth Stokoe, In Case of Emergency, Order Pizza: An Urgent Case of Action Formation and Recognition, in: *Proceedings of the 1st International Conference on Conversational User Interfaces*, 1–2. CUI '19, New York 2019.

6   Philipp Kirschthaler/Martin Porcheron/Joel E. Fischer, What Can I Say? Effects of Discoverability in VUIs on Task Performance and User Experience, in: *Proceedings of the 2nd Conference on Conversational User Interfaces*, New York 2020, 1–9.

7   Harold Garfinkel, Ethnomethodology's Program: Working out Durkeim's Aphorism, Lanham 2002.

8   Harold Garfinkel/Harvey L. Sacks, On Formal Structures of Practical Actions, in: John C. McKinney/Edward A. Tiryakian (eds.), *Theoretical Sociology: Perspectives and Developments*, New York 1970, 338–66.

of interactions.[9] Our study mainly explored the ways in which people adopted to the device by altering their speech style to VUI-speak.

2. Related Work: VUIs and Conversation

Natural language processing and understanding (NLP/NLU) research on the role of the spoken word relating to "conversation" has largely been conducted using computational linguistics, dialogue systems, computational sociolinguistics, Gricean pragmatics, cognitive semantics, and psycholinguistics.[10] General findings from these approaches aim at addressing design issues regarding, among other things, word choice and the most relevant types of mappings between words and their pragmatic functions.[11] A variety of action and coding schemes have been developed based on interpretations of speech act theory, resulting in formalized predicate calculus and plan-based models of dialogue.[12] Although ordinary developments of interfaces and apps with VUIs are not coded while, for example, Grice's book *Studies in the Way of Words*[13] is lying on the table as a guide for the programmer, Google's developer guidelines[14] are nonetheless explicitly based on an introduction to conversation based on Grice's cooperative principle and its maxims. One crucial challenge is that this approach treats the concept of *social action* as a matter of intentions and cognitive functions, omitting the intrinsic sequential organization from which such actions emerge. Consequently, these approaches in general (e.g., cognitive linguistics, speech act theory, Gricean implicature) result in fundamental constraints on how VUIs should deal with the pervasive problems of action formation and action recognition in speech.[15]

---

9   Lucy Suchman, *Plans and Situated Actions: The Problem of Human-Machine Communication*, Cambridge 1987; Graham Button et al., *Computers, Minds and Conduct*, Cambridge 1995; Joel E. Fischer et al., Beyond 'Same Time, Same Place': Introduction to the Special Issue on Collocated Interaction, in: *Human–Computer Interaction* 33 (5–6/2018), 305–10; Christian Heath/Jon Hindmarsh/Poul Luff, *Video in Qualitative Research*, London 2009.

10 Alexander Clark/Chris Fox/Shalom Lappin, *The Handbook of Computational Linguistics and Natural Language Processing*, Malden/Oxford 2013.

11 Chris Cummins/Jan P. de Ruiter, Computational Approaches to the Pragmatics Problem, in: *Language and Linguistics Compass* 8 (4/2014), 133–43.

12 Daniel Jurafsky/James H. Martin, *Speech and Language Processing: An Introduction to Natural Language Processing, Computational Linguistics and Speech Recognition* (1st edition), Upper Saddle River 2000.

13 Paul Grice, *Studies in the Way of Words*, Cambridge 1989.

14 Google, Designguidelines, Conversation Design, Designguidelines, 2020. URL: https://designguidelines.withgoogle.com/conversation/conversation-design/welcome.html# [last accessed: August 15, 2023].

15 Cf. Albert/Housley/Stokoe, In Case of Emergency, Order Pizza.

In this paper, we build on the work of EM/CA approaches to HCI, NLU, and VUIs. Although it is not a dominant theoretical approach, EM/CA researchers have, over the years, consistently provided the HCI community with novel insights and, more recently, have also furnished insights into human–robot interaction (HRI.)[16] EM/CA studies related to VUIs have focused on the devices' embeddedness in everyday life, such as in homes[17] or public environments.[18] Our contribution forms a part of this small but growing branch of studies dealing with VUIs in natural settings.

2.1   EM/CA Research on Turn-Taking and Repairing Actions When Talking With a VUI

We built the concept of VUI-speak based on basic theoretical understandings of how human interaction is organized in sequences that, from the outset, comprise base sequences residing as adjacency pairs (e.g., question–answer/request–response).[19] A sequence entails one or more verbal actions designed to be recognizable to the recipient; that is, a question is an action formatted in a such a way as to be recognizable to the recipient *as* a question, and this is observably the case when a recipient responds to the first pair part with an answer.[20] Problems arise in conversation when people produce "wrong" turns or words, which are then displayed and repaired in the conversation.[21]

EM/CA research has specifically dealt with participants' interactional work to produce actions that ensure progressivity. Fischer et al.[22] have shown how a lack of

---

16   Karola Pitsch et al., Interactional Dynamics in User Groups: Answering a Robot's Question in Adult-Child Constellations, in: *Proceedings of the 5th International Conference on Human Agent Interaction*, New York 2017, 393–97; Hannah Pelikan/Mathias Broth/Leelo Keevallik, Are You Sad, Cozmo?' How Humans Make Sense of a Home Robot's Emotion Displays, in: *HRI '20: Proceedings of the 2020 ACM/IEEE International Conference on Human-Robot Interaction*, New York 2020, 461–470.

17   Porcheron et al., Voice Interfaces in Everyday Life.

18   Martin Porcheron/Joel E. Fischer/Sarah Sharples, 'Do Animals Have Accents?': Talking with Agents in Multi-Party Conversation, in: *Proceedings of the 2017 ACM Conference on Computer Supported Cooperative Work and Social Computing*, New York 2017, 207–19.

19   Harvey L. Sacks/Emmanuel A. Schegloff/Gail Jefferson, A Simplest Systematics for the Organization of Turn-Taking for Conversation, in: *Language* 50 (4/1974), 696–735; Stephen C. Levinson, On the Human 'Interaction Engine', in: N.J. Enfield, Stephen C.Levinson (eds.), *Roots of Human Sociality: Culture, Cognition and Interaction*, New York 2006.

20   Emmanuel A. Schegloff, Sequence Organization in Interaction: A Primer in Conversation Analysis, New York 2007; Stephen C. Levinson, Action Formation and Ascription, in: Jack Sidnell/Tanya Stivers (eds.), *The Handbook of Conversation Analysis*, Oxford 2012, 101–30.

21   Emmanuel A. Schegloff/Gail Jefferson/Harvey L. Sacks, The Preference for Self-Correction in the Organization of Repair in Conversation, in: *Language* 53 (2/1977), 361–82.

22   Joel E. Fischer et al., Progressivity for Voice Interface Design, in: *Proceedings of the 1st International Conference on Conversational User Interfaces*, 1–8. CUI '19. New York 2019.

second pair responses, accounts, and what they refer to as "non-answer responses" may not only impede the overall activity but also require that the participants figure out what went wrong. Pelikan and Broth[23] have demonstrated how users deal with the limitations of automatic speech recognition by adapting their requests to the robot's limited perceptive abilities as these become apparent in the interaction. Though those scholars have found that adaptation is noticeably trouble-free for the user, they also found, in line with Fischer et al., that it is the user who does the social interactional work and that the sequential coordination of turns-at-talk remains troublesome.

In human-to-human face-to-face (F2F) interaction, a lack of a verbal response would not necessarily pose a problem since humans employ multiple resources in their efforts to create mutual understanding and common ground by using, for example, visual resources.[24] In VUI interaction, however, the participants rely solely on speech. This means that the participants must figure out "what to do next" if the device does not react as expected and does not understand natural language. In other words, the participants cannot rely on the VUI to support them reflectively in their progress towards successful completion of the ongoing action, and they conversely adapt to the machine algorithm. Hence, repair, that is, participants' sequential management of interactional "trouble," is a common activity in VUI interaction.[25]

A key distinction within EM/CA research on repair is between initiating and producing the repair solution. Research has shown that there is a preference for self-initiated repair,[26] where the speaker who produced the "trouble source" also produces the repair. However, repair can also be other-initiated, where another participant prompts the speaker to self-repair or even provides a repair solution themselves. Although the VUI does not provide the same kind of repair solutions as humans would, it still initiates repair and provides suggestions for other actions. However, as we will show in the analysis, as a consequence of the VUI's indexical referential limitation to only concern the last turn of a sequence, speakers' self-repair does not lead to intersubjectivity and progression of the ongoing activity as in human–human interaction but rather leads to a lapse.

One practice for reducing failure and repair in VUI interaction is to change the way of speaking to the device by switching from using ordinary language to pro-

23   Hannah R.M. Pelikan/Mathias Broth, Why That Nao?: How Humans Adapt to a Conventional Humanoid Robot in Taking Turns-at-Talk, in: *Proceedings of the 2016 CHI Conference on Human Factors in Computing Systems*, New York 2016, 4921–32.

24   Lorenza Mondada, The Local Constitution of Multimodal Resources for Social Interaction, in: *Journal of Pragmatics*, A body of resources – CA studies of social conduct 65 (2014), 137–56.

25   David Frohlich/Paul Drewl/Andrew Monk, Management of Repair in Human-Computer Interaction, in: *Human–Computer Interaction* 9 (3–4/1994), 385–425.

26   Schegloff/Jefferson/Sacks, The Preference for Self-Correction.

ducing commands in sequences. We are not suggesting that this is a new finding. Different types of institutional interactions entail specialized turn-taking systems, for example, in a courtroom, in journalistic interviews, and in similar settings where specific formal procedures are supposed to be followed.[27] In more informal settings, participants have been observed as adopting a specific way of designing turns and choosing words in the pursuit of achieving shared understanding when interacting with elderly people who may have a disability.[28] When analyzing institutional interaction, regardless of the level of formality, the task, then, extends from identifying sequential patterns to also detailing their use in accomplishing the institutional activity.[29] Turn production is always recipient designed, one way or another. For the speaker, the task is to produce speech that embeds recognizable actions for the participants. This relates to the design of action or "action formation."

2.2 EM/CA Research on Action Formation and Directives

Action formation is a key issue in VUI interaction, as Albert et al.[30] have shown. We contribute to this research by showing how participants, through an unfolding sequence, work their way towards successful action formation by formulating it as VUI-speak. Whereas current EM/CA research treats participants' action formation when talking to the VUI as requests,[31] it seems more appropriate to characterize them as commands, that is, a branch of directives. Searle described directives as having an illocutionary force.[32] However, this is a broad category. Conversation analysis has shed light on how directives differ from requests. Requests are, according to Curl and Drew,[33] actions in which one participant *asks* another to do something. According to Craven and Potter,[34] directives are actions where one participant *tells* another to do something. Directives produced as orders and commands are mostly

---

27   Paul Drew/John Heritage, *Talk at Work: Interaction in Institutional Settings*, Cambridge 1992; Mie Femø Nielsen et al., Interactional Functions of Invoking Procedure in Institutional Settings, in: *Journal of Pragmatics* 44 (11/2012), 1457–73.

28   Susan Kemper, Elderspeak: Speech Accommodations to Older Adults, in: *Aging, Neuropsychology, and Cognition* 1 (1/1994), 17–28; Elisabeth D. Kristiansen/Gitte Rasmussen/Elisabeth Muth Andersen, Practices for Making Residents' Wishes Fit Institutional Constraints: A Case of Manipulation in Dementia Care, in: *Logopedics Phoniatrics Vocology* 44 (1/2019), 7–13.

29   Ilkka Arminen, *Institutional Interaction: Studies of Talk at Work*, New York 2005.

30   Albert/Housley/Stokoe, In Case of Emergency, Order Pizza.

31   Stuart Reeves, Conversation Considered Harmful?, in: *Proceedings of the 1st International Conference on Conversational User Interfaces*, New York 2019, 1–3.

32   John Searle, *Expression and Meaning. Studies in the Theory of Speech Acts*, Cambridge 1979.

33   Traci S. Curl/Paul Drew, Contingency and Action: A Comparison of Two Forms of Requesting, in: *Research on Language and Social Interaction* 41 (2/2008), 129–53.

34   Alexandra Craven/Jonathan Potter, Directives: Entitlement and Contingency in Action, in: *Discourse Studies* 12 (4/2010), 419–442.

directed at children or animals or are used in the military or under similar circumstances. All of these should be considered institutional interactions where the participants' institutional or professional identities are made relevant in the ongoing task-oriented activity.[35] The distinct formal pattern in institutional interaction then becomes the participants' way of recognizing the type of interaction, as well as their way of organizing the accomplishment of a practical institutional task.[36] Directives are far less common in symmetrical ordinary conversation because there appears to be a conversational preference for producing requests, which involves fine-tuning the turns to uphold the recipients' face. As task-oriented asymmetrical interaction with a fixed preformatted turn-taking system, VUI-speak is to be considered institutional in its nature because it is preformatted. Consequently, we will present and discuss how a characteristic of VUI-speak is related to ways of designing a workable directive turn.

In the study reported in this paper, we investigated how visually impaired people (VIP) interact with VUIs. The benefits of VUIs for VIP are significant, as they enable vocal rather than graphical interfaces. Prior research has mainly focused on specialized assistive technology, without accounting for understandings of the practical accomplishments of technology-in-use. Few previous EM/CA studies have dealt with VIP's actual use of technology,[37] and only one recent paper by Reyes-Cruz et al.[38] reports on studies of VIP interacting with VUI, showing that ambient noise and environmental issues affect the interaction.

3. Method and Data

VIP may benefit, in particular, from using digital assistants for practical tasks that otherwise would require visual orientation and visual interfaces. Although the current study's findings and theoretical contribution have implications for the broader understanding and development of VUIs, as a user group, VIP constitute a perspicuous case. Due to their lack of one sense (sight), VIP are more reliant on hearing, sounds, and language. Garfinkel called marginal cases involving VIP "natural exper-

35   Drew/Heritage, Talk at Work.

36   Arminen, Institutional Interaction.

37   Brian Due et al., Technology Enhanced Vision in Blind and Visually Impaired Individuals. Synoptik Foundation Research Project, in: *Circd Working Papers in Social Interaction* 3 (1/2017), 1–31; Gisela Reyes-Cruz/Joel E. Fischer/Stuart Reeves, Reframing Disability as Competency: Unpacking Everyday Technology Practices of People with Visual Impairments, in: *Proceedings of the 2020 CHI Conference on Human Factors in Computing Systems*, New York 2020, 1–13.

38   Gisela Reyes-Cruz/Joel Fischer/Stuart Reeves, An Ethnographic Study of Visual Impairments for Voice User Interface Design, in: ArXiv:1904.06123 [Cs], 2019.

iments."[39] Such studies can inform and challenge basic theory because they may reveal the taken-for-granted aspects of ordinary practices.[40]

The data used in this study derived from an ongoing (2020–2023) project investigating blind and visually impaired persons' use of new mainstream and experimental technologies, predominantly focusing on computer vision and NLP, for example, VUIs and smartphone apps. We recruited seven blind and visually impaired adults comprising two cohabiting couples and three people living alone (or with non-participating roommates) to capture their natural use of the Google Home Assistant in their familiar home environments. Several cameras were strategically placed around the living areas and near the VUI devices to record the participants' everyday practices as they interacted with the VUIs. Thus far, over a six-month period, we have collected approximately 30 hours of recordings from the participants' homes.

We examined the 30 hours of video recordings, including a total of 38 sequences of VUI interaction, and identified an overall occurrence of what we came to label *VUI-speak*. Through detailed EM/CA analysis of the data, we identified three characteristics of VUI-speak and initiated a collection supporting our initial findings. To make our case easy to follow, in this article, we will use examples from a single setting and provide only one example in each of the three categories.

The excerpt examined in this chapter derived from data recorded on July 1, 2019, depicting a 58-year-old visually impaired adult interacting with the VUI and the researcher in a natural setting. The participant informally chatted with the researcher, listened to music from Spotify through the VUI speakers, and eventually tuned into a radio show on YouTube that was streamed via TV. In this setting, and throughout our data corpus, the Google Home device was set to Danish.

The scene begins in the participant's living room (Fig. 1), where the participant and the researcher are sitting and drinking coffee, while discussing their favorite music, podcasts, and radio shows, and chatting about how Google Home can be used to easily find and play different types of media.

39   Anne W. Rawls/Kevin A. Whitehead/Waverly Duck (eds.), *Black Lives Matter—Ethnomethodological and Conversation Analytic Studies of Race and Systemic Racism in Everyday Interaction*, New York 2020, see 8–9.

40   Mairian Corker/Tom Shakespeare, *Disability/Postmodernity: Embodying Disability Theory*, New York 2002.

Fig. 1: *Graphic overview of setting. Camera 2 (C2) is placed by the Google Home Assistant (dot).*

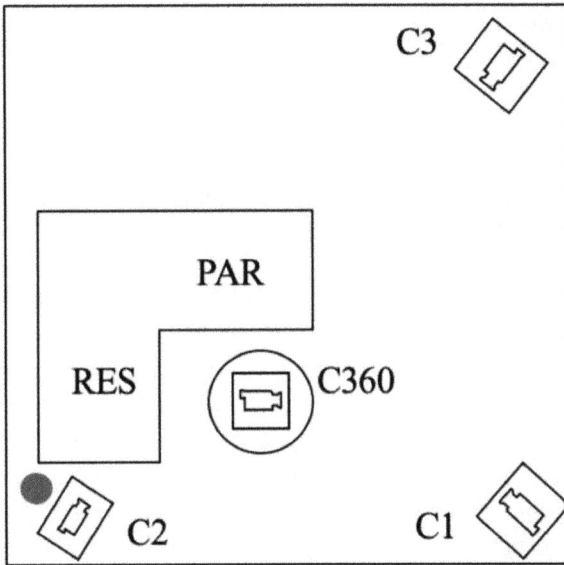

The participant initially uses the VUI to play music by Johnny Cash through the speakers, but after some conversational exchanges with the researcher about their favorite radio shows, the participant decides that they should listen to a radio show instead. The overall activity comprises two independent modifications, as shown in Figure 2: changing "media" from the VUI speakers to TV and changing "platform" from Spotify to YouTube.

Fig. 2: *Two-part task of changing from Johnny Cash to The Short Radio Show.*

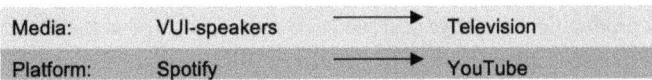

Media:	VUI-speakers	⟶	Television
Platform:	Spotify	⟶	YouTube

4. Analysis

VUI-speak is recognizable as a participant practice in and through the following three characteristics, which will be outlined in greater detail in the following pages: VUI-speak (1) occurs at the third position embedded within a five-part sequential structure, 2) comprises a change of action formation design, and 3) has a non-pausing application-oriented turn design.

4.1   VUI-Speak at the Third Position Embedded Within a Five-Part Sequential Structure

In the following paragraphs, the sequential structure of VUI-speak is described as a five-part base sequence. This structure was identified in the data, but it can also be seen as an extension of what we already know about basic sequences in HCI, per Arminen's 2005 work, which resulted in a three-step model of a basic HCI sequence, based on data obtained from a human participant who was observed browsing on a visual printer display; each step consists of human action and a resulting change on the display. Thus, Arminen's steps can be considered as three coherent adjacency pairs, each consisting of an activity and a response. We further expanded the model's structure by dividing each step into turns based on the fact that the participants in the interaction treat VUI "thinking-bleep-sounds" as turns that display acknowledgment of the prior turn.

For a participant to succeed at their VUI-directed activity, they must recognize the structure and produce specific types of actions fitted to that structure. (This pertains specifically to the third position, where the directive is produced.)

The five positions are illustrated in the following excerpt (Fig. 3). The first four positions recur repeatedly: (1) A participant produces an awake call, which is the first pair part of the pre-sequence, summoning the VUI (occasionally, with a self-initiated repair and upgrade, e.g., Line 1). Next, (2) the VUI responds with a readiness display as the second pair part. These two pairs bear resemblance to a summons and to picking up the receiver in a telephone conversation. (3) The participant responds to this by producing a directive designed as a command, (4) and the VUI then produces a response display, followed by (5) either a VUI action, an account from either the participant or the VUI, or silence, and finally, a rerun of the sequence. Let us unfold this in more detail based on the excerpt shown in Figure 3.

Fig. 3: Transcript line 1–20, three reruns of the five-part sequence.

1.	Awake call	→ 01 PAR:	hey google (3.1) HEY GOOGLE
2.	Readiness display	→ 02 VUI:	((bleep↓)) (0.7)
3.	Directive	→ 03 PAR:	sluk for johnny cash
			turn off johnny cash
4.	Response display	→ 04 VUI:	((bleep↑)) (0.5)
		05 VUI:	det fangede jeg ik (1.0)
5.	Action or accounting	→	*i did not get that* (1.0)
		06 VUI:	((music and youtube continues))
1.	Awake call	→ 07 PAR:	HEY GOOGLE
2.	Readiness display	→ 08 VUI:	((bleep↓))
3.	Directive	→ 09 PAR:	STOP musik (1.4)
4.	Response display	→ 10 VUI:	((bleep↑))
		11 VUI:	((youtube turns off; music
5.	Action or accounting	→	continues)) (5.0)
		12 PAR:	så stopped det der
Expanded sequence			*then it stopped there*
		13 RES:	ja (4.0)
			yes
1.	Awake call	→ 14 PAR:	hey google
2.	Readiness display	→ 15 VUI:	((bleep↓)) (1.3)
3.	Directive	→ 16 PAR:	STOP johnny cash (1.3)
4.	Response display	→ 17 VUI:	((bleep↑)) (.)
		18 VUI:	det fangede jeg ik
5.	Action or accounting	→ 19	*i didn't get that*
		20 VUI:	((johnny cash continues))

The awake call (Fig. 3, Position 1, Line 1), "Hey, Google," is a standard command given to activate the Google Home Assistant. In this case, it is first produced in line with the participant's normal style of speaking.[41] An ensuing pause of 3.1 seconds with no response from the VUI prompts the participant to repair the awake call and reissue it at a higher volume. We consider this to be an inserted expansion of the first position within the sequence. Thus, in the first production of the awake call, we note how the participant has had to adapt their way of speaking to the device (i.e., using VUI-speak). When the VUI registers the awake call, it responds immediately with an audible four-note bleep (Fig. 3, Position 2, Line 2) produced with an audible downward intonation. The audible readiness and response displays are accompanied by a visual light on the top of the Google Home speaker. The displays can be deselected if the user wishes to do so, in which case they will only consist of a visual display, giving the sequential Positions 2 and 4 a far more subtle and thus far less interactive character. However, since the study participants are all either visually impaired or blind, the preferred setting is always with an audible display. The audible sound (accompanied by a visual light) is recognizable as displaying readiness to receive the next action. In this case, the participant produces a command in the next turn: "Turn off

---

41   Erving Goffman, *Frame Analysis: An Essay on the Organization of Experience*, New York 1974.

Johnny Cash" (Fig. 3, Position 3, Line 3).[42] In the excerpt provided in Fig. 3, the VUI then produces a response display comprising a bleep (Fig. 3, Position 4, Line 4) with an upward intonation, indicating that the command has been registered. The two forms of the VUI's minimal responses differ audibly. The difference between what we call the *readiness display* (Fig. 3, Position 2) and the *response display* (Fig. 3, Position 4) is not just recognizable in and through its sequential position but also with regard to its action formation: Whereas an audible readiness display is composed with a falling intonation, a response display has a rising intonation. Figure 4 shows an audible analysis of the two forms designed to display the VUI's different stances of being ready to receive or process the directive, respectively. The differences are hearable during interaction.

Fig. 4: A prosodic analysis of line 15 (readiness display, falling intonation) + line 17 (response display, rising intonation). Produced via Praat.

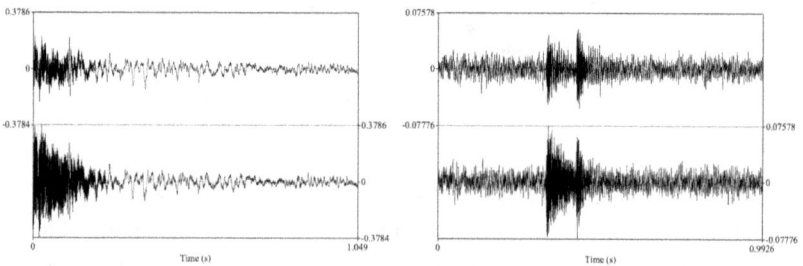

After the response display in the fourth position, one of two things will typically happen: Either (a) the command effects the required VUI action (or it keeps doing what it is doing, e.g., Fig. 3, Line 11), or (b) there is a reply from the VUI, explaining difficulties with the command (e.g., Fig. 3, Line 5). When the VUI does not produce an accounting explanation for why the directive is not working (Fig. 3, Position 5), the participant will typically account for the trouble in an expanded sequence (e.g., Fig. 3, Lines 12–13) and change the recipient designed action formation, thus producing a repaired rerun of the five-part structure. This is evident in Figure 3, which shows three reruns due to the difficulty of prompting the VUI to produce the required action. In cases where the VUI does not understand the directive, as in Line

---

42   Analysis of our corpus data has revealed that participants normally need to produce a directive with a VUI-recognizable action formation within an approximately 1.2(/3) second time frame after the readiness display, or the VUI interrupts the interaction. Figure 3 shows cases of 0.7 seconds (Line 2), 0 seconds (Line 7), and 1.3 seconds (Line 15), respectively, before a directive is produced. An example of a troublesome too-late third position directive is provided in Figure 8, to which we will return later.

5 of Figure 3, its account is a simple "I did not get that," without further elaboration of what went wrong or what the participant needs to do next. The VUI closes the sequence as well as its readiness to receive further directives, forcing the participant to initiate a rerun of the five-part sequence. By essentially shutting down after its response, the VUI deprives the user of the opportunity to repair by indexically referring to the context just built up. Thus, the VUI does not recognize changes in action formation design as repair, leaving the participant to re-summon it to effect repair. An illustration of the sequential structure is provided in Figure 5.

Fig. 5: The sequential organization of the five-part structure that circulates.

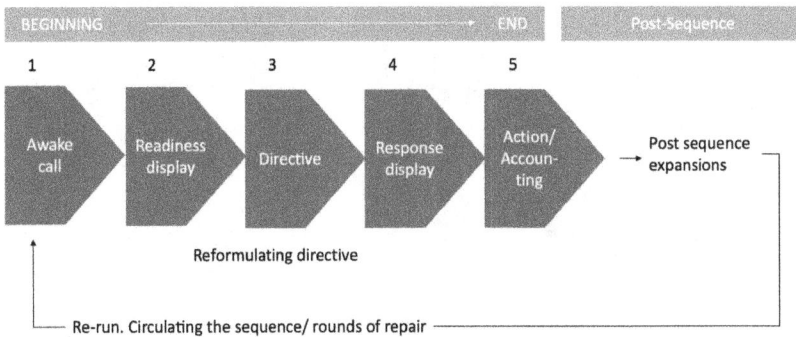

4.2   VUI-Speak Comprising a Change of Action Formation Design

We have shown the occurrence of a five-part sequential structure in a VUI (Google Home). Now, we will present another structural and recurrent phenomenon that we also consider to be constitutive of VUI-speak. We suggest calling this a members' practice of *change of action formation in VUI interaction*. Eliciting a response from the VUI requires a command. Commands are, as discussed earlier, part of directive actions. However, it is not just a matter of producing commands (instead of requests) that is the issue for participants but rather *what* specific commands to produce. A recurrent theme is that participants repair and change the lexical turn constructional units and their sequential placement. Figure 6 shows a related excerpt, as continued from Figure 3.

Fig. 6: transcript of lines 27–47

1.	Awake call	→27	PAR:	hey google
2.	Readiness display	→28	VUI:	((bleep↓))
3.	Directive	→29	PAR:	genoptag video (1.2)
				resume video
4.	Response display	→30	VUI:	((bleep↑))
5.	Action or accounting	31	VUI:	noget gik galt (0.5)
	→			*something went wrong*
		32		prøv igen når du er
				try again when you are
		33		klar til det (1.1)
				ready
				((A few lines omitted))
1.	Awake call	→34	PAR:	hey google
2.	Readiness display	→35	VUI:	((bleep↓))
3.	Directive	→36	PAR:	spil (1.2) [via]
				play through
4.	Response display	→37	VUI:	[((bleep↑))]
5.	Action or accounting	→38	PAR:	bvarbvava >det er det
				it is that
		39		man [ska] vide hvad det e:r
				you need to know what it is
		40	RES:	[m:m]
		41	PAR:	man vil sige< før
				you want to say before
		42	VUI:	((johnny cash starts again))
Expanded sequence		43	PAR:	°>man begynder på det
				you start saying what
		44		man vil sige<°
				you are going to say
		45	RES:	ja (3.6)
				yes
		46	PAR:	°man slipper >han er svær
				you can't he is hard
		47		at slippe af med hva<°
				to get rid off huh

In the case provided, the TV and the VUI speakers are both playing simultaneously, and the participant is trying to turn off the music playing from the VUI speakers, so that they can listen exclusively to the video playing on the TV. Figure 7 illustrates the process through seven reruns of the five-part structure. Although the VUI does not contribute to the progression of the interaction by suggesting what to do next and instead leaves it up to the participant to figure this out, the VUI response does, nonetheless, reveal a small insight into how the algorithm works after each attempt. It is this situated and sequential progression that the participant orients towards and builds upon in these sequences.

In Line 9 of Figure 3, the participant commands, "Stop music," resulting in the VUI stopping the wrong device (the TV instead of Spotify). The command fails, as the desired action has not been achieved, and the participant responds by changing the design of the subsequent action formation. Something does in fact stop when the command "Stop" (Fig. 3, Line 11) is given, so it can be inferred that "Stop" is a workable command for something. It can also be inferred that the word "music" is too broad a signifier when several platforms are in use. These interpretations do not represent our attempts at a cognitive analysis, but it is notable that, in the next turns, the participant reuses the word "Stop" (Fig. 3, Line 16) but modifies "music" to the more specific phrase "Johnny Cash." However, this wording does not ensure compliance either, as evidenced by the fact that the fourth position is occupied by the VUI's accounting explanation for an understanding problem (Fig. 3, Line 18). Hence, through an exclusion process, the participant determines that "Johnny Cash" is the unintelligible factor. Since we can follow his interpretive process as observable action, we notice that now, the participant, instead of specifying the musician's name, specifies the app in the third rerun of the five-part structure by giving the command "Stop Spotify" (Fig. 6, Line 25), with which the VUI complies by causing the music to stop playing. The first part of the overall task is, at this point, accomplished, owing to the use of the command "Stop" combined with the specification of the app from which the unwanted music originated, namely, "Spotify" (instead of the musician's name, "Johnny Cash," used to describe the actual music). *VUI-speak* is thus evident in the way the participant changes his word choice in a process that adapts to the VUI's responsiveness.

The participant then initiates the second part of the overall task: playing the radio show. He does this by referring to a previous action (playing the radio show via a YouTube video streamed on TV while simultaneously playing music from the speakers [not in this transcript]) and commanding, "Resume video" (Fig. 6, Line 29). The VUI does not comply with this command; it accounts with the uninformative explanation "Something went wrong" (Fig. 6, Line 30), and initiates a repair by inviting the participant to "Try again when you are ready," which implies that the VUI is not at fault and that the participant will need to provide the repair solution. Based on this VUI response, the participant knows that the ongoing interaction is not indexically related to previous turns or tasks. In other words, every directive needs to be recipient designed for the VUI with reference to immediately prior and/or ongoing VUI actions. Consequently, "Resume" is an unworkable command.

In the next rerun of the five-part structure, the participant produces the command "Play" (Fig. 6, Line 36). However, he hesitates for 1.2 seconds before producing the directive, and resultantly, the VUI responds by turning on the music once again. Though the command failed, the participant gains two new insights: "Play," like "Stop," is a minimal command that will trigger immediate action with no initial account; further, it is essential not to pause beyond 1.2 seconds of maximum silence during a command (see Section 4.3). To turn off the music again, the participant reiterates, "Stop music," a command with which the VUI complies without a further account as there is only one type of media playing at this point. As shown in Figure 7, the participant continues the process until a successful directive is produced (Fig. 8, Line 50). We will show the detailed organization of workable action formation in the following section.

Fig. 7: A specific model of the change of action formation, showing how PAR rephrases his directive according to VUI action.

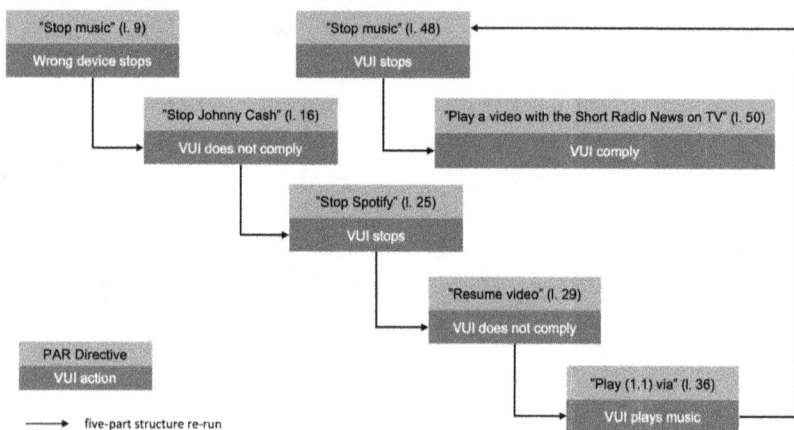

4.3   VUI-Speak as Non-Pausing Application-Oriented Turn Design

The participant needs to produce a directive in the third position (within the five-part structure) with a *non-pausing and application-oriented turn design*. It is evident from Line 33 of Figure 6 that because the participant did not produce a VUI recognizable action within the time frame of 1.0–1.3 seconds, the VUI treated the turn as completed and produced a response display (a bleep; Fig. 6, Line 35). Hence, pauses play a pivotal role in interacting with VUIs. Whereas earlier AI technologies such as Google Glass produced long pauses (over four seconds)[43] and were beyond the normal one-second standard maximum period of silence in conversation, as Jefferson described,[44] Google Home follows the normal maximum silence between turns more precisely. Still, the issue in natural interaction is *not* a specific between-turn pause duration rule but rather the contextual and pragmatic understanding of *when* a pause represents problems or repairables related to transition relevant places and when they represent a speaker's resource for formulating turns.[45] For instance, in

---

43   Brian L. Due, The Social Construction of a Glasshole: Google Glass and Multiactivity in Social Interaction, in: *PsychNology* 13 (2–3/2015), 149–78.

44   Gail Jefferson, Notes on a Possible Metric Which Provides for a 'standard Maximum' Silence of Approximately One Second in Conversation, in: Gail Jefferson (ed.), *Tilburg Papers in Language and Literature*, Tilburg 1983, 1–83.

45   Harvey L. Sacks/Emmanuel A. Schegloff/Gail Jefferson, A Simplest Systematics for the Organization of Turn-Taking for Conversation, in: *Language* 50 (4/1974), 696–735; Elliott M. Hoey, Lapses: How People Arrive at, and Deal With, Discontinuities in Talk, in: *Research on Language and Social Interaction* 48 (4/2015), 430–53.

Line 50 of Figure 8, the participant formulates the directive as an extended command with no intra-turn pausing and adheres to the other previously acquired situated insights of using minimal commands and references to specific apps. Given that Google Home is presumably designed according to the principles for micro-interactions, framed by Miller in 1968[46] as being a few seconds of waiting time and a minimal amount of user commands[47] (see also the concluding discussion on Gricean principles in Section 5), the interactional task for the participant would appear to be to use VUI-speak by producing VUI recognizable actions within a VUI recognizable turn design.

Fig. 8: Non-pausing application-oriented turn designed.

50	PAR:	<SPIL	EN	VIDEO	MED	DEN KORTE RADIOAVIS	PÅ	T:V:>
		<PLAY	A	VIDEO	WITH	THE SHORT RADIO NEWS	ON	T:V:>
		command		app	reference to	content	reference to	device

In Figure 8, Line 50, we can observe that the participant uses a minimal command ("Play"), a specified application ("a video"), a reference ("with"), specified content ("The Short Radio News"), another reference ("on"), and a designated device ("TV"). Although this is recognizable ordinary and natural language, what the participant is really doing is treating natural language as programming language, as follows: command app reference content reference device. These are produced in staccato bursts with articulated pronunciation at a high volume with an emphasis on the first minimal command ("Play"). This is easily comprehensible language. It is recognizable as a turn and also appears to adhere to grammatical rules. It makes sense as natural language. At the same time, the turn is designed as a directive and is successful precisely because it adheres to the device's computational infrastructure. This form of directive would represent an extreme case in natural conversation. However, it is successful in this particular setting because it is recognizable VUI-speak.

---

46   Robert B. Miller, Response Time in Man-Computer Conversational Transactions, in: *Proceedings of the Fall Joint Computer Conference*, New York 1968, 267–77.

47   Thad Starner, Project Glass: An Extension of the Self, in: *IEEE Pervasive Computing* 12 (2/2013), 14–16.

5. Conclusion

We have, in this chapter, suggested terminology for the types of practices in which participants engage when naturally interacting with VUIs, namely Google Home in this particular case. To recap, this terminology is centered around the occurrence of "members' methods" [48] for doing *VUI-speak*, that is, ways of adapting their forms of speech to fit the system. This relates both to producing actions at the right sequential position (in regard to Google Home, that would be the third position) and with an appropriate action design. We call these workable action formations *VUI recognizable actions*. The system evidently does not understand actions that indexically reference larger contexts (e.g., "Resume video"; Fig. 6, Line 29), and hence, this chapter supports research such as Albert et al.'s, [49] criticizing VUIs' currently limited understanding of action formation. Whereas other forms of sequential structures that differ from the five-part structure may exist in interactions with other types of VUIs (besides Google Home), there is currently, per definition, always a slot in VUI interaction where a directive is supposed to be produced. Accordingly, the general lesson to be drawn from this study is that a directive needs to be produced at a specific sequential moment, at a specific pace, and with a specific action formation design.

We have shown in the analysis that to facilitate self-repair, the participant needs to "start over" by beginning with summoning the VUI, and this needs to happen each and every time. The correction then becomes more comprehensive than a repair as it exceeds a momentary departure from the main part of the interaction and entails closing the sequence after initiating repair. In that sense, other-initiated repair by the VUI cannot be considered to be cooperative behavior and thus just a momentary side sequence; [50] rather, it is equivalent to an error display encouraging the participant to adjust their way of speech accordingly.

The final successful directive in the studied case is characterized by a distinctive design we have proposed calling *non-pausing application-oriented turn design*, and we have argued that this is clearly not conversation, despite its composition of recognizable and grammatical language, but rather VUI-speak. It is therefore interesting to note that people's natural interactions adapt to the system processually in a way that adheres to something resembling Gricean cooperative principles. [51] In-

---

48   Harold Garfinkel/Harvey Sacks, On Formal Structures of Practical Actions*, in: Harold Garfinkel, *Ethnomethodological Studies of Work*, London 1986, 160–94.

49   Albert/Housley/Stokoe, In Case of Emergency, Order Pizza.

50   Mark Dingemanse/N.J. Enfield, Other-Initiated Repair across Languages: Towards a Typology of Conversational Structures, in: *Open Linguistics* (1/2015), 96–118.

51   H. Paul Grice, Logic and Conversation, in: P. Cole/J. Morgan (eds.), *Syntax and Semantics*, Volume 3: Speech Acts, University Park 1975, 41–58.

deed, this is no coincidence, as the entire conversational architecture of Google's VUI is designed precisely based on these maxims of quality, quantity relevance, and manner, as prescribed in Google's design guidelines.[52] However, although Gricean implicature provides a solid theoretical understanding of cooperative principles in interaction, and they can be readily accepted as types of members' resources *in practice*,[53] complications arise if they are treated as static concepts and applied as such in VUI design. Successful design based on static Gricean principles results in a static VUI system, which works as principles and maxims (as a programmable language) but *lacks* a conversational interface. Designing with an understanding of sequencing and action formation as situated accomplishments is recommended instead.

As Reeves et al.[54] have also pointed out, describing HCI as "conversational" is confusing. The term "conversational" is used to describe systems that display more humanlike characteristics and support the use of spontaneous natural language in contrast to systems that require a more restricted form of user input such as single words or short phrases.[55] However, as EM/CA research, such as Reeves et al., has shown, the current versions of VUIs only work as request–response systems. The current chapter supports use of the descriptor *"conversation-sensitive design."* [56] We have, in line with this, shown that conversational interfaces are *not* really conversational,[57] and we have offered terminology to describe the actual practices: practices of *VUI-speak* and orientation to a *five-part structure*, and practices of *non-pausing application-oriented turn design* for the successful accomplishment of a command.

In future studies aimed at VUI design, it would be worthwhile to explore a connection between Gricean implicature and conversation analysis given that, for humans, the same words in an utterance can contain different implicatures according to each recipient's perspective, and this is observable in the kind of response the recipient produces. VUIs should be able to adapt gradually to the participant's actions— not the other way around.

---

52   Google, Conversation Design.

53   Jack Bilmes, Ethnomethodology, Culture, and Implicature: Toward an Empirical Pragmatics, in: *Pragmatics* 3 (4/1993), 387–409; Paul Drew, The Interface between Pragmatics and Conversation Analysis, in: Cornelia Llie/Neal R. Norrick (eds.), *Pragmatics and Its Interfaces*, Amsterdam; Philadelphia 2018, 59–85.

54   Stuart Reeves/Martin Porcheron/Joel Fischer, 'This Is Not What We Wanted': Designing for Conversation with Voice Interfaces, in: *Interactions* 26 (1/2018), 46–51.

55   Michael McTear/Zoraida Callejas/David Griol, *The Conversational Interface: Talking to Smart Devices*, Cham 2016.

56   Reeves, Conversation Considered Harmful?.

57   Porcheron et al., Voice Interfaces in Everyday Life.

Bibliography

Albert, Saul/Housley, William/Stokoe, Elizabeth, In Case of Emergency, Order Pizza: An Urgent Case of Action Formation and Recognition, in: *Proceedings of the 1st International Conference on Conversational User Interfaces*, 1–2. CUI '19, New York 2019.

Arminen, Ilkka, *Institutional Interaction: Studies of Talk at Work*, New York 2005.

Bilmes, Jack, Ethnomethodology, Culture, and Implicature: Toward an Empirical Pragmatics, in: *Pragmatics 3* (4/1993), 387–409.

Breazeal, Cynthia/Dautenhahn, Kerstin/Kanda, Takayuki, Social Robotics, in: Bruno Siciliano/Oussama Khatib (eds.), *Springer Handbook of Robotics*, Berlin; Heidelberg 2016, 1935–71.

Button, Graham/Coulter, Jeff/Lee, John/Sharrock, Wes, *Computers, Minds and Conduct*, Cambridge 1995.

Clark, Alexander/Fox, Chris/Lappin, Shalom, *The Handbook of Computational Linguistics and Natural Language Processing*, Malden/Oxford 2013.

Corker, Mairian/Shakespeare, Tom, *Disability/Postmodernity: Embodying Disability Theory*, New York 2002.

Craven, Alexandra/Potter Jonathan, Directives: Entitlement and Contingency in Action, in: *Discourse Studies 12* (4/2010), 419–442.

Cummins, Chris/de Ruiter, Jan P., Computational Approaches to the Pragmatics Problem, in: *Language and Linguistics Compass 8* (4/2014), 133–43.

Curl, Traci S./Drew, Paul, Contingency and Action: A Comparison of Two Forms of Requesting, in: *Research on Language and Social Interaction 41* (2/2008), 129–53.

developers.google.com, Conversation Design, Google Developers, 2021. URL: https ://developers.google.com/assistant/conversation-design/learn-about-convers ation [last accessed: August 15, 2023].

Dingemanse, Mark/Enfield, N.J., Other-Initiated Repair across Languages: Towards a Typology of Conversational Structures, in: *Open Linguistics* (1/2015), 96–118.

Drew, Paul, The Interface between Pragmatics and Conversation Analysis, in: Cornelia Llie/Neal R. Norrick (eds.), *Pragmatics and Its Interfaces*, Amsterdam; Philadelphia 2018, 59–85.

Drew, Paul/Heritage, John, *Talk at Work: Interaction in Institutional Settings*, Cambridge 1992.

Due, Brian L., The Social Construction of a Glasshole: Google Glass and Multiactivity in Social Interaction, in: *PsychNology 13* (2–3/2015), 149–78.

Due, Brian L./Kupers, Ron/Lange, Simon/Ptito, Maurice, Technology Enhanced Vision in Blind and Visually Impaired Individuals. Synoptik Foundation Research Project, in: *Circd Working Papers in Social Interaction 3* (1/2017), 1–31.

Fischer, Joel E./Reeves, Stuart/Brown, Barry/Lucero, Andrés, Beyond 'Same Time, Same Place': Introduction to the Special Issue on Collocated Interaction, in: *Human–Computer Interaction* 33 (5–6/2018), 305–10.

Fischer, Joel E./Reeves, Stuart/Porcheron, Martin/Sikveland, Rein O., Progressivity for Voice Interface Design, in: *Proceedings of the 1st International Conference on Conversational User Interfaces*, 1–8. CUI '19. New York 2019.

Frohlich, David/Drew, Paul/Monk, Andrew, Management of Repair in Human-Computer Interaction, in: *Human–Computer Interaction* 9 (3–4/1994), 385–425.

Garfinkel, Harold, *Ethnomethodology's Program: Working out Durkeim's Aphorism*, Lanham 2002.

Garfinkel, Harold/Sacks, Harvey, On Formal Structures of Practical Actions*, in: Harold Garfinkel, *Ethnomethodological Studies of Work*, London 1986, 160–94.

Garfinkel, Harold/Sacks, Harvey L., On Formal Structures of Practical Actions, in: John C. McKinney/Edward A. Tiryakian (eds.), *Theoretical Sociology: Perspectives and Developments*, New York 1970, 338–66.

Goffman, Erving, *Frame Analysis: An Essay on the Organization of Experience*, New York 1974.

Google, Designguidelines, Conversation Design, Designguidelines, 2020. URL: https://designguidelines.withgoogle.com/conversation/conversation-design/welcome.html# [August 15, 2023].

Grice, H. Paul, Logic and Conversation, in: P. Cole/J. Morgan (eds.), *Syntax and Semantics. Volume 3: Speech Acts*, University Park 1975, 41–58.

Grice, Paul, *Studies in the Way of Words*, Cambridge 1989.

Heath, Christian/Hindmarsh, Jon/Luff, Poul, *Video in Qualitative Research*, London 2009.

Hoey, Elliott M., Lapses: How People Arrive at, and Deal With, Discontinuities in Talk, in: *Research on Language and Social Interaction* 48 (4/2015), 430–53.

Jefferson, Gail, Notes on a Possible Metric Which Provides for a 'standard Maximum' Silence of Approximately One Second in Conversation, in: Gail Jefferson (ed.), *Tilburg Papers in Language and Literature*, Tilburg 1983, 1–83.

Jurafsky, Daniel/Martin, James H., *Speech and Language Processing: An Introduction to Natural Language Processing, Computational Linguistics and Speech Recognition* (1st edition), Upper Saddle River 2000.

Kemper, Susan, Elderspeak: Speech Accommodations to Older Adults, in: *Aging, Neuropsychology, and Cognition* 1 (1/1994), 17–28.

Kirschthaler, Philipp/ Porcheron, Martin/Fischer, Joel E., What Can I Say? Effects of Discoverability in VUIs on Task Performance and User Experience, in: *Proceedings of the 2nd Conference on Conversational User Interfaces*, New York 2020, 1–9.

Kristiansen, Elisabeth Dalby/Rasmussen, Gitte/Muth Andersen, Elisabeth, Practices for Making Residents' Wishes Fit Institutional Constraints: A Case of Manipulation in Dementia Care, in: *Logopedics Phoniatrics Vocology* 44 (1/2019), 7–13.

Levinson, S.C., On the Human 'Interaction Engine', in: N.J. Enfield, S.C.Levinson (eds.), *Roots of Human Sociality: Culture, Cognition and Interaction*, New York 2006.

Levinson, Stephen C, Action Formation and Ascription, in: Jack Sidnell/Tanya Stivers (eds.), *The Handbook of Conversation Analysis*, Oxford 2012, 101–30.

McTear, Michael/Callejas, Zoraida/Griol, David, *The Conversational Interface: Talking to Smart Devices*, Cham 2016.

Miller, Robert B., Response Time in Man-Computer Conversational Transactions, in: *Proceedings of the Fall Joint Computer Conference*, New York 1968, 267–77.

Mondada, Lorenza, The Local Constitution of Multimodal Resources for Social Interaction, in: *Journal of Pragmatics, A body of resources – CA studies of social conduct* 65 (2014), 137–56.

Nielsen, Mie Femø/Beck Nielsen, Søren/Gravengaard, Gitte/Due, Brian L., Interactional Functions of Invoking Procedure in Institutional Settings, in: *Journal of Pragmatics* 44 (11/2012), 1457–73.

Pearl, Cathy, *Designing Voice User Interfaces: Principles of Conversational Experiences* (1st edition), Beijing 2017.

Pelikan, Hannah/Broth, Mathias/Keevallik Leelo, Are You Sad, Cozmo?' How Humans Make Sense of a Home Robot's Emotion Displays, in: *HRI '20: Proceedings of the 2020 ACM/IEEE International Conference on Human-Robot Interaction*, New York 2020, 461–470.

Pelikan, Hannah R.M./Broth, Mathias, Why That Nao?: How Humans Adapt to a Conventional Humanoid Robot in Taking Turns-at-Talk, in: *Proceedings of the 2016 CHI Conference on Human Factors in Computing Systems*, New York 2016, 4921–32.

Pitsch, Karola/Gehle, Raphaela/Dankert, Timo/Wrede, Sebastian, Interactional Dynamics in User Groups: Answering a Robot's Question in Adult-Child Constellations, in: *Proceedings of the 5th International Conference on Human Agent Interaction*, New York 2017, 393–97.

Porcheron, Martin/Fischer, Joel E./Reeves, Stuart/Sharples, Sarah, Voice Interfaces in Everyday Life, in: *Proceedings of the 2018 CHI Conference on Human Factors in Computing Systems* 640, New York 2018, 1–12.

Porcheron, Martin/Fischer, Joel E./Sharples, Sarah, 'Do Animals Have Accents?': Talking with Agents in Multi-Party Conversation, in: *Proceedings of the 2017 ACM Conference on Computer Supported Cooperative Work and Social Computing*, New York 2017, 207–19.

Rawls, Anne Warfield/Whitehead, Kevin A./Duck, Waverly (eds.), *Black Lives Matter – Ethnomethodological and Conversation Analytic Studies of Race and Systemic Racism in Everyday Interaction*, New York 2020.

Reeves, Stuart, Conversation Considered Harmful?, in: *Proceedings of the 1st International Conference on Conversational User Interfaces*, New York 2019, 1–3.

Reeves, Stuart/Porcheron, Martin/Fischer, Joel, 'This Is Not What We Wanted': Designing for Conversation with Voice Interfaces, in: *Interactions* 26 (1/2018), 46–51.

Reyes-Cruz, Gisela/Fischer, Joel E./Reeves, Stuart, Reframing Disability as Competency: Unpacking Everyday Technology Practices of People with Visual Impairments, in: *Proceedings of the 2020 CHI Conference on Human Factors in Computing Systems*, New York 2020, 1–13.

Reyes-Cruz, Gisela/Fischer, Joel/Reeves, Stuart, An Ethnographic Study of Visual Impairments for Voice User Interface Design, in: ArXiv:1904.06123 [Cs], April 12, 2019. http://arxiv.org/abs/1904.06123.

Sacks, Harvey L./Schegloff, Emmanuel A./Jefferson, Gail, A Simplest Systematics for the Organization of Turn-Taking for Conversation, in: *Language* 50 (4/1974), 696–735.

Schegloff, Emmanuel A., *Sequence Organization in Interaction: A Primer in Conversation Analysis*, New York 2007.

Schegloff, Emmanuel A./Jefferson, Gail/Sacks, Harvey L., The Preference for Self-Correction in the Organization of Repair in Conversation, in: *Language* 53 (2/1977), 361–82.

Searle, John, *Expression and Meaning. Studies in the Theory of Speech Acts*, Cambridge 1979.

Starner, Thad, Project Glass: An Extension of the Self, in: *IEEE Pervasive Computing* 12 (2/2013), 14–16.

Statista, Google Home Global Shipments 2016–2025, Statista, 2020. URL: https://www.statista.com/statistics/1022722/worldwide-google-home-unit-shipment / [last accessed: August 15, 2023].

Suchman, Lucy, *Plans and Situated Actions: The Problem of Human-Machine Communication*, Cambridge 1987.

Doing Family on Unfamiliar Terrain
The Constitution and Contestation of Kinship Between Two Humans, Two Cats, and a Voice Assistant[1]

Miriam Lind

Abstract *The concept of family has undergone tremendous changes throughout the last century, shifting from a firmly established social institution that is to a much more loosely defined social structure that is done. This relates both to questions of who is part of a family and which practices are involved in the doing of families. One aspect of this shift is the increasing inclusion of nonhuman entities in family units, that is, the understanding of pets as family members and the care practices involved in this social construction.*

With the rise of voice user interfaces that enable verbal interaction between humans and machines and the subsequently increasing presence of "talking machines," that is, voice assistants, in family homes, questions have arisen regarding these entities' social status within families. This article reports an autoethnographic study on the practices of doing family in a household comprising two humans, two cats, and a voice assistant to illuminate how inclusion in and exclusion from the family is socially constructed through interaction.

1. Introduction

"With everything Echo can do, it's really become part of the family." This sentence concluded the four-minute long "Introducing Amazon Echo" video that Amazon used to present its new smart speaker Echo and its inbuilt voice assistance system Alexa to the world.[2] The video depicts the ideal traditional family—white suburban middle class heterosexual parents with three children[3]—who acquaint the audience with the functions of this new "family member," which encompass knowledge-focused question–response sequences, playing music, offering recipes and writing

---

1   This research was funded by the German Research Foundation (DFG) through the CRC 1482 "Human Differentiation".
2   URL: https://www.youtube.com/watch?v=zmhcPKKt7gw [last accessed: August 15, 2023].
3   Cf. Thao Phan, Amazon Echo and the aesthetics of whiteness, in: *Catalyst: Feminism, Theory, Technoscience* 5 (1/2019), 1–39.

shopping lists, and use as a communicative resource[4] in sibling banter. Although Amazon's Superbowl 2020 commercial for its smart speaker portrays Alexa as "downgraded" to a mere service function in the spotless designer home of American celebrity Ellen DeGeneres and her wife, Portia de Rossi, the assistance system is nevertheless attributed such an important role in their lives that "life before Alexa," as the video calls it, seems unimaginable.[5]

These commercial promises of voice assistants' seamless positioning within the most private parts of the social world seem at odds with the amount of research focusing on the multitude of ways in which spoken language-based human–machine interaction fails and that it is primarily the human interlocutor who adapts their communicative behavior to be more "machine-friendly."[6] Although most of this research is conducted in experimental environments, current studies increasingly use "in the wild" settings to gain deeper insights into the ways human and artificial interlocutors interact, through video analysis and/or interviews.[7] Interestingly, studies focusing on the communicative behavior between humans and voice assistants in family settings routinely fail to explicitly define what they mean by "family," seemingly using the term overall synonymously with heterosexual couples with

---

4   Cf. Deborah Tannen, 'Talking the Dog: Framing Pets as Interactional Resources in Family Discourse, in: *Research on Language and Social Interaction* 37 (4/2004), 399–420.

5   Cf. Sascha Dickel/Miriam Schmidt-Jüngst, Gleiche Menschen, ungleiche Maschinen. Die Humandifferenzierung digitaler Assistenzsysteme und ihrer Nutzer:innen in der Werbung, in: Dilek Dizdar et al. (eds.), *Humandifferenzierung. Disziplinäre Perspektiven und empirische Sondierungen*, Velbrück 2021, 342–367.

6   E.g., Erin Beneteau et al., Alexis, Communication breakdowns between families and Alexa, in: *CHI '19: Proceedings of the 2019 CHI Conference on Human Factors in Computing Systems*, Glasgow 2019, 4–9; Jiepu Jiang/Wei Jeng/Daqing He, How do users respond to voice input errors?: Lexical and Phonetic Query Reformulation in Voice Search, in: *Proceedings of the 36th International ACM SIGIr Conference on Research and Development in Information Retrieval*, 2013, 143–152; Manja Lohse et al., "Try something else!" – When users change their discursive behavior in human-robot interaction, in: *2008 IEEE International Conference on Robotics and Automation*, Pasadena 2008, 3481–3486; Ewa Luger/Abigail Sellen, "Like having a really bad PA": The gulf between user expectation and experience of conversational agents, in: *Proceedings of the 2016 CHI Conference on Human Factors in Computing Systems*, 2016, 5286–5297., Jamie Pearson et al., Adaptive language behavior in HCI: How expectations and beliefs about a system affect users' word choice, in: *Proceedings of the SIGCHI Conference on Human Factors in Computing Systems*, 2006, 1177–1180.

7   E.g., Diana Beirl/Yvonne Rogers/Nicola Yuill, Using voice assistant skills in family Life, in: *Proceedings of the 13th international conference on computer supported collaborative learning – A wide lens: Combining embodied, enactive, extended, and embedded learning in collaborative settings*, CSCL 2019, 96–103., Beneteau et al., Alexis, Communication breakdowns between families and Alexa; Alisha Pradhan/Leah Findlater/Amanda Lazar, "Phantom Friend" or "Just a Box with Information": Personification and Ontological Categorization of Smart Speaker-based Voice Assistants by Older Adults, in: *PACM on Human-Computer Interaction* 3 (2019), 1–21.

children (although Beneteau et al.'s study[8] included several single parents living together with one or more children).[9] Families, in the context of human–machine interaction, appear as reified institutions that ostensibly automatically emerge with a child's birth. These approaches to human–machine interaction in family settings miss the opportunity to perspectivize the practices involved in the performative construction of human family in the presence of an "artificial companion."[10] Contrastingly, there is rich scholarly work on posthuman and interspecies practices of doing family in the context of human–pet relations.[11]

This paper presents an autoethnographic pilot study on the doing and undoing of family and kinship between humans, cats, and Amazon's Alexa in a household positioned on the margins of traditional understandings of family: a queer German–British couple, their two adopted cats, and the newest member of the household, a fourth-generation Amazon Echo with the voice assistance system Alexa. Based on the logs Amazon's Alexa program automatically stores during and following interactions with the device and "reflexive investigation,"[12] this study analyzed interaction and communicative behavior in the household and sought to ascertain the ways in which the artificial companion was included and excluded in the practices of doing family, how technical obstacles and communication breakdowns affected these practices, and how human–machine interaction has become embedded in human beliefs and attitudes towards family, technology, and interaction. This study thus provides a critical approach to human–machine interaction "in the wild" and examines how the introduction of voice assistants into the privacy of homes, as well as into family systems, impacts our understanding of communication and the "fragile institutionalisation"[13] of family. To achieve this goal, this

---

8   Cf. Beneteau et al., Alexis, Communication breakdowns between families and Alexa.

9   E.g., Beirl/Rogers/Yuill, Using voice assistant skills in family Life; Beneteau et al., Alexis, Communication breakdowns between families and Alexa, Olivia K. Richards, Family-centered exploration of the benefits and burdens of digital home assistants, in: *Extended Abstracts of the 2019 CHI conference on human factors in computing systems*, New York 2019, 1–6.

10   Andreas Hepp, Artificial companions, social bots and work bots: communicative robots as research objects of media and communication studies, in: *Media, Culture & Society* 42 (7–8/2020), 1410–1426.

11   E.g., Leslie Irvine/Laurent Cilia, More-than-human families: Pets, people, and practices in multispecies households, *Sociology Compass* 11 (2/2017), 1–13; Melissa Laing, On being posthuman in human spaces: critical posthumanist social work with interspecies families, in: *International Journal of Sociology and Social Policy* 41 (3/4/2021), 361–375., Nicole Owens/Liz Grauerholz, Interspecies parenting: How pet parents construct their roles, in: *Humanity & Society* 43 (2/2018), 1–24.

12   Garance Maréchal, Autoethnography, in: Albert J. Mills/Elden Wiebe (eds.), *Encyclopedia of case study research* (vol. 1), Thousand Oaks 2010, 43–45.

13   Kurt Lüscher, Familie – Von der Institution zu einer fragilen Institutionalisierung, in: *Recht der Jugend und des Bildungswesens* 56 (2/2008), 120–125.

paper starts by providing an overview of what doing family entails in human and interspecies families, as well as by reviewing the literature on human interaction with voice assistance systems. The research methodology is then discussed before the data are presented and analyzed.

2. Doing Family

2.1 Doing human families: From institutions to fragile institutionalization

In recent decades, concepts of *family* in the social sciences have shifted away from seeing them as institutions that *are* and towards understandings of family as social structures that *are done*.[14] Although law and legislation largely continue to reify this conceptualization of family as organizational units with the care, education, and socialization of children at their core,[15] other academic disciplines, as well as the media and large parts of society, have adapted to a more nuanced perspective on the diverse, multifaceted reality of family forms in the 21st century: Family has "turned into a 'project', for which one has to do something—for it to occur, to be preserved and for the desired quality of togetherness to emerge."[16] Family is thus not—at least, not anymore—a self-evident institution that exists as an unquestioned, reified thing in the world but is rather an accumulation of practices and decision-making processes through which individuals negotiate whether and how they are and do family. Despite the acknowledgement of this de-institutionalization of family, many scholars continue to understand family as indispensably linked to reproduction and the upbringing of children; Lüscher, for example, writes that instead of focusing "on the institution [of family] as such," one needs to pay attention to "the tasks that need to be performed." Lüscher defines these tasks in the context of family "as the human necessity of caring for, nurturing and raising [...] children."[17] Similarly, Hertz characterizes the contemporary family as a "remarkably elastic institution, with women and their children"[18] at their heart, who flexibly move between households occupied by a wide range of other members of the household who can but do not need to be understood as parts of this family. Although children do play a crucial role in many

---

14   Karin Jurczyk, Doing Family – der Practical Turn der Familienwissenschaften, in: Anja Steinfach/Marina Hennig/Oliver Arránz Becker (eds.), *Familie im Fokus der Wissenschaft*, Wiesbaden 2017, 117–138.

15   E.g. Ferdinand Kerschner, Bürgerliches Recht. Band V: Familienrecht, Wien/New York 2010.

16   Jurczyk, Doing Family, 117.—When German sources are quoted, they are presented in translation. All translations are mine.

17   Lüscher, Familie – Von der Institution zu einer fragilen Institutionalisierung, 121.

18   Rosanna Hertz, Talking About "Doing" Family, in: *Journal of Marriage and Family* 68 (2006), 796–799, see 797.

definitions of family, even more important is the aspect of caretaking and nurturing among family members, as "nurturing, a central and daily aspect of family, sustains group life."[19] This becomes particularly evident when Jurczyk writes that

> family is a historically and culturally highly changeable system of care-oriented and emotion-based relationships between generations and genders which are geared towards reliable commonality, but which need to be (re)produced and can change throughout family development and family constellations.[20]

The author further points out three core areas in which our Western understanding of family has changed: (1) Consanguinity, marriage, and the traditional division of labor play a smaller role in our lives; (2) families can be better described as multi-local networks than as localized to one specific shared household; and (3) families are less commonly founded on predefined traditional values; that is, family is not a preexisting given resource in society but rather a social practice dependent on constant performance and negotiation.[21] This performativity of family takes three main forms, the first of which Jurczyk calls balance management and defines as the organizational and logistic tasks conducted by family members that are necessary to make family livable in everyday life.[22] The second form of doing family, according to Jurczyk, is the construction of commonality, that is, those interactive processes that perform family as a shared, complete entity.[23] Constructing this sense of commonality takes place through reciprocal reference to each other, acts that are performed together, and members' own definition as a family and the semiotic construction of belonging, often through linguistic practices (e.g., family-exclusive pet names or a family-specific register of meaning). Jurczyk describes the third form of doing family as "displaying family,"[24] a practice that is directed outwards and is particularly relevant for those families that do not fit within the traditional idea of family and therefore require legitimization.

Blackstone is one of the very few who specifically investigated the practices involved in the doing of family in childless families.[25] That families without children have so rarely been addressed in research is, according to her, a product of the sociocultural reduction of women to the role of mothers: "For if childbearing is culturally understand [sic] to be at the 'core of women's experience' (Hird 2003, 6), then it

---

19   Hertz, Talking About "Doing" Family, 797.
20   Jurczyk, Doing Family, 117–18.
21   Cf. Jurczyk, Doing Family, 124.
22   Cf. Jurczyk, Doing Family, 129.
23   Jurczyk, Doing Family, 129.
24   Jurczyk, Doing Family, 129.
25   Cf. Amy Blackstone, Doing Family Without Having Kids, in: *Sociology Compass* 8 (1/2014), 52–62.

makes sense that family scholars and others might overlook the familial forms and experiences of women who are not mothers."[26]

Blackstone suggests that instead of trying to "reconcile definitions of the family" with the lived realities of families without children, an alternative might be to consider "the functions that families help societies meet"[27] in order to understand what counts as family. She describes these societal functions of the family as "(i) providing emotional and sexual companionship for members; (ii) [making] economic provisions for members; (iii) providing a home to members; and (iv) [facilitating] biological and social reproduction."[28] Of these functions, the first three indicate that the existence of children within the family unit is not necessary, and Blackstone's extended understanding of reproduction to mean not only biological but also social reproduction provides a basis to acknowledge the contributions childless families can make to society's functioning. This social reproduction includes all forms of care work and education and the passing on of knowledge, values, and ideas within a community that are necessary for the community's sustainable continuation.[29] The author further points out that adults who do not have children are more engaged in charitable and voluntary work than parents and thus play a crucial role in their communities. This illustrates that families without children "do family" in similar ways as families with children: They provide for each other emotionally, sexually, and economically, and they regularly engage in forms of social reproduction that maintain their society's wellbeing and sustainability. The author concludes her thoughts on childless families with the suggestion that family units might stretch beyond the human sphere:

> Researchers might also consider childless families that include members beyond a couple. This could mean more than two adults but, as recent research and popular discourse demonstrate, non-human animals such as pets may also play a significant role in the families of childless adults.[30]

These human–animal families will be investigated in more detail in the next section.

---

26   Blackstone, Doing Family Without Having Kids, 56.
27   Blackstone, Doing Family Without Having Kids, 53.
28   Blackstone, Doing Family Without Having Kids, 53.
29   Cf. Blackstone, Doing Family Without Having Kids, 57.
30   Blackstone, Doing Family Without Having Kids, 58.

2.2   Doing interspecies families

The re-figuration of the relationship between humans and their nonhuman companions as family members is a recent development in Western societies that dissolves the ontological boundary between the human and the nonhuman animal.[31]
That the idea of an interspecies family is no longer unthinkable is "one of the more intriguing changes"[32] in our conceptualizations of family that have emerged throughout the last century. The societal shifts that have contributed to the emergence of
interspecies families are, according to these two authors,

> delayed age at marriage, the deinstitutionalization of marriage (e.g., increased
> cohabitation and [an] increase in [the] numbers of individuals who remain
> unmarried), [the] increase in child-free/childless women, experimentation
> with varying family arrangements as a matter of choice, and new pathways to
> parenthood.[33]

However, it is not only changes in the human understanding of family that have led
to the changed positionality of pets in human–animal relationships; the perception
of animals and their categorization as threat, livestock, or pet have changed dramatically throughout human history, and locating them even just within the household
is a recent development.[34] Whether pets can be part of family is a question closely
related to their personhood. Sanders argues that an "animal's personhood is an interactive accomplishment" that is negotiated in its relationship with humans and
continues to define an animal's personhood as contingent on their "perspective and
feeling [being] knowable; interaction [being] predictable; and the shared relationship [providing] an experience of closeness, warmth, and pleasure."[35] The main distinction Sanders draws between human–human and human–animal relationships
is the contingency of the former.[36] Some authors have therefore claimed that "pets
occupy a liminal status: domestic, but not human; family member, but still 'other,'"[37]
or that they "stand at the intersection of kind and kin."[38] In these attempts to define

---

31   Cf. Rebekah Fox, Animal behaviours, post-human lives: everyday negotiations of the an-
     imal-human divide in pet-keeping, in: *Social & Cultural Geography* 7 (4/2006), 525–537.
32   Owens/Grauerholz, Interspecies parenting, 2.
33   Owens/Grauerholz, Interspecies parenting, 3.
34   Cf. Irvine/Cilia, More-than-human families.
35   Clinton R. Sanders, Actions Speak Louder than Words: Close Relationships between Hu-
     mans and Nonhuman Animals, in: *Symbolic Interaction* 26 (3/2003), 405–426, see 418.
36   Cf. Sanders, Actions Speak Louder than Words.
37   Irvine/Cilia, More-than-human families, 3.
38   Marc Shell, The Family Pet, in: *Representations* 15 (1986), 121–153.

pets' positioning in the human sphere, a clear distinction is made between "the animal" in general and "the pet" in particular, as only the latter are considered potential participants in the familial construct. An animal can be considered a pet, according to Irvine and Cilia, when they are given individual names, are allowed in the house, and are not considered food.[39] The reference to naming practices as playing an important role in the consideration of an animal as a pet and therefore, by extension, in the attribution of personhood highlights the importance of linguistic signs in the construction of the social world. Language similarly has a crucial function in the construction of interspecies families, especially by extending family terminology to human–pet relationships (e.g., calling a pet one's *child* or *fur baby*[40] and referring to oneself as a *pet parent* or even the animal's *mom* or *dad*)[41] by naming pets with personal names similar to those suitable for human children[42] or by using pet-directed speech that shows high degrees of similarity to infant-directed speech (also known as *motherese* or *baby talk*).[43] Further practices that position a pet firmly in the family are their inclusion in holiday celebrations (e.g., pet birthdays, pet weddings, or pets' receipt of Christmas gifts)[44] and their inclusion in (human) obituaries and birth announcements[45] or the creation of obituaries and eulogies for the pets themselves.[46] From a legal perspective, questions regarding "custody" of shared pets in

---

39   Cf. Irvine/Cilia, More-than-human families.

40   Cf. Jessica Greenebaum, It's a Dog's Life: Elevating Status from Pet to "Fur Baby" at Yappy Hour, in: *Society and Animals* 12 (2/2004), 117–135.

41   Cf. Owens/Grauerholz, Interspecies parenting.

42   Cf. Damaris Nübling, Tiernamen als Spiegel der Mensch-Tier-Beziehung. Ein erster Einblick in die Zoonomastik, in: *Sprachreport* 31 (2/2015), 1–7; Ralph Slovenko, The Human/Companion Animal Bond and the Anthropomorphizing and Naming of Pets, in: *Med Law* 2 (1983), 277–283.

43   E.g. Tobey Ben-Aderet et al., Dog-directed Speech: why do we use it and do dogs pay attention to it?, in: *Proc. R. Soc. B.* 284 (2017), 1–7; Denis Burnham, Denis/Christine Kitamura/Uté Vollmer-Conna, What's New, Pussycat? On Talking to Babies and Animals, in: *Science* 296 (5572/2002), 1435.

44   E.g. Nancy M. Ridgway et al., Does excessive buying for self relate to spending on pets?, in: *Journal of Business Research* 61 (5/2008), 392–396.

45   E.g. Angelika Linke/Jan Anward, Familienmitglied ‚Vofflan'. Zur sprachlichen Konzeptualisierung von Haustieren als Familienmitglieder. Eine namenpragmatische Miniatur anhand von Daten aus der schwedischen Tages- und Wochenpresse, in: Antje Dammel/Damaris Nübling/Mirjam Schmuck (eds.), *Tiernamen – Zoonyme. Band 1: Haustiere*, Heidelberg 2015, 77–96; Cindy C. Wilson et al., Companion Animals in Obituaries: An Exploratory Study, in: *Anthrozoös: A multidisciplinary journal on the interactions of people and animals* 26 (2/2013), 227–236.

46   E.g. Jill R.D. MacKay/Janice Moore/Felicity Huntingford, Characterizing the Data in Online Companion-dog Obituaries to Assess Their Usefulness as a Source of Information about Human-Animal Bonds, in: *Anthrozoös: A multidisciplinary journal on the interactions of people and animals* 29 (3/2016), 431–440; Jane Rennard/Linda Greening, Linda/Jane M. Williams, In Praise of Dead Pets: An Investigation into the Content and Function of Human-Style

divorces[47] or pets' rights to inherit[48] are points of increasing discussion. Irvine and Cilia further point out that pets regularly "reshape everyday family practices"[49] because they need care and nurture, humans share intimate relations with them, they share the household, and they impact household routines, for example, by waking family members up and urging them to play, to go on walks, or to provide food. In these aspects of human–pet relationships, the similarities to the forms of human doing family outlined in the previous section become apparent: providing emotional companionship, economically providing for household members, and providing a home. A further important feature of these interspecies families is that they are and will as long as they exist be inherently asymmetrical: Similar to a baby or a toddler, pets are dependent on their usually adult human owners who have to provide for them, and communicatively, they often fulfil similar functions in interactions as infants do.[50] Moreover, whereas human children grow up eventually and might then be able to interact with their parents within a more symmetrized power dynamic and even potentially care and provide for their parents at a later stage in life, pets do not leave this infant-like state of dependency: They will, for their entire lives, need to be cared for, and although we regularly talk to them, they do not understand what is said beyond trained commands, and they cannot talk back.[51]

A point that will become particularly relevant when discussing voice assistants' social status in households was raised by Irvine and Cilia regarding the role pets play in the raising and aging of human children: Pets are seen as important participants in the socialization of children, who speak to them and confide in them more often than they speak to or confide in their human siblings.[52] Additionally, interacting with and caring for pets is assumed to be beneficial for children's acquisition of kindness, as treating animals with kindness is often generalized to acting kindly towards other humans.[53] Irvine and Cilia further state that

---

Pet Eulogies, in: Anthrozoös: A multidisciplinary journal on the interactions of people and animals 32 (6/2019), 769–783.

47 E.g. Lacy L. Shuffield, Pet Parents – Fighting Tooth and Paw for Custody: Whether Louisiana Courts Should Recognize Companion Animals as more than Property, in: Southern University Law Review 37 (1/2009), 101–125.

48 E.g. Frances H. Foster, Should Pets Inherit?, in: Florida Law Review 63 (4/2011), 801–856.

49 Irvine/Cilia, More-than-human families, 4.

50 Cf. Jörg R. Bergmann, Haustiere als kommunikative Ressourcen, in: Hans-Georg Soeffner (ed.), Kultur und Alltag, Göttingen 1988, 299–312; Tannen, Talking the Dog.

51 Cf. Bergmann, Haustiere als kommunikative Ressourcen.

52 Cf. Irvine/Cilia, More-than-human families, 5.

53 Cf. Irvine/Cilia, More-than-human families; Janine C. Muldoon et al., Promoting a "duty of care" towards animals among children and young people: a literature review and findings from initial research to inform the development of interventions, Department for Environment, Food and Rural Affairs, London 2009.

[a]cts of kindness and cruelty illustrate how children, pets, and parents actively engage in doing family. Children's close relationships with pets, their communication with their pets, and their tendency to define pets as siblings reveal both their conception of a flexible human-animal boundary and how it intersects with their ideas about family.[54]

The idea that children's interactions with nonhuman entities in their household especially impact the ways in which they interact with fellow humans—in other words, communication with nonhumans can be an influential factor in a child's upbringing—is prominently discussed in research on the interaction between humans and voice assistance systems. Nevertheless, in the context of human–machine interaction, it has not been discussed whether and how this impacts the doing of family and the social positioning of artificial entities within the family, as will be shown in the following section.

3. Interacting With Voice Assistance Systems

Voice assistance systems are, simply put, "software agents that can interpret human speech and respond via synthesized voices."[55] The novelty of these systems lies in the fact that "voice has become the primary interface"[56] in interaction with them, thus positioning them as potential interlocutors in conversation. This promise of conversationality, an ability that has, so far, been understood as strictly human, makes them prone to anthropomorphization[57] and the attribution of agency, personhood, and social roles.[58] Although voice assistance technology was popularized as a feature on smartphones, beginning with the self-proclaimed "humble personal assistant" Siri on the Apple iPhone 4S in 2011, it has now found a permanent place in people's homes in the form of smart speakers integrated into the smart home's permanent

---

54   Irvine/Cilia, More-than-human families, 6–7.

55   Matthew B. Hoy, Alexa, Siri, Cortana, and More: An Introduction to Voice Assistants, in: *Medical Reference Services Quarterly* 37 (1/2018), 81–88, see 81.

56   Martin Porcheron/Joel E. Fischer/Stuart Reeves/Sarah Sharples, Voice Interfaces in Everyday Life, in: *Proceedings of the 2018 CHI Conference on Human Factors in Computing Systems (CHI '18)*. ACM, New York 2018, 1–12.

57   E.g. Juliana Schroeder/Nicholas Epley, Mistaking Minds and Machines: How Speech Affects Dehumanization and Anthropomorphism, in: *Journal of Experimental Psychology* 145 (11/2016), 1427–1437.

58   E.g. Pradhan/Findlater/Lazar, "Phantom Friend" or "Just a Box with Information"; Amanda Purington, Amanda/Jessie G. Taft/Shruti Sannon/Natalya N. Bazarova, "Alexa is my new BFF": Social Roles, User Satisfaction, and Personification of the Amazon Echo, in: *CHI EA '17: Proceedings of the 2017 CHI Conference Extended Abstracts on Human Factors in Computing Systems*, 2017, 2853–2859.

interconnectedness. This setting in the privacy of people's homes seems to have instantaneously raised public and academic interest in the ways families interact with voice assistants, as evidenced by the scholarly articles "Communication Breakdowns Between Families and Alexa,"[59] "Using Voice Assistant Skills in Family Life,"[60] and "Family-Centered Exploration of the Benefits and Burdens of Digital Home Assistants,"[61] which are just some of the recent studies taking empirical approaches to the use of voice assistance systems in people's homes. Families appear in these studies as the primary occupants of homes but are simultaneously left undefined: Beneteau et al. recruited "10 diverse families" for their study, "who represent a wide-spectrum of family life" and had, at the time, "at least one child between the ages of four and 17 living in the home,"[62] with the number of adults considered as family members ranging between one and three. Who these adult family members were (e.g., grown-up children, parents, partners, grandparents, etc.), how these families were constituted beyond the members of one household, and on what basis families were defined are not apparent in the study. Beirl, Rogers, and Yuill conducted an "in-the-wild study [...] in six family's homes" and "[a]ll had children in the age group of 2–13 years."[63] Through a table summary, the reader is informed about the number of children and the parental situation (3 x mother and father, 2 x mother, 1 x 2 mothers), but the paper provides no further information regarding potential additional family members. Richards' study took a different approach: She based her analysis on online reviews of the Amazon Echo that were "filtered for relevant content using words such as *kid, child, son, daughter,* and *grand* (for 'grandchild')."[64] Similar to the previously mentioned studies, in this research, *family* seems to be solely predicated on the existence of under-age children living in the same household as at least one parent. This is all the more surprising as all three studies were driven by interest in aspects of family practices: Beneteau et al. used a framework of "family collaboration and joint media engagement"[65] to analyze how families collaboratively repair communication breakdowns in interactions with Alexa; Beirl, Rogers, and Yuill assessed how "families learn [...] new [Alexa] skills and appropriate them into their life" by exploring "different types of interactions [...] that facilitated family cohesion, bond-

---

59   Beneteau et al., Alexis, Communication breakdowns between families and Alexa.

60   Beirl/Rogers/Yuill, Using voice assistant skills in family Life.

61   Olivia K. Richards, Family-centered exploration of the benefits and burdens of digital home assistants, in: *Extended Abstracts of the 2019 CHI conference on human factors in computing systems,* New York 2019, 1–6.

62   Beneteau et al., Alexis, Communication breakdowns between families and Alexa, 4.

63   Beirl/Rogers/Yuill, Using voice assistant skills in family Life, 98.

64   Richards, Family-centered exploration of the benefits and burdens of digital home assistants, 3.

65   Beneteau et al., Alexis, Communication breakdowns between families and Alexa, 6.

ing and empathy";[66] and Richards investigated "the burdens and benefits of digital home assistants to parents with children,"[67] a phrasing that leaves the reader wondering whether there are parents without children and if so, whether they would have been included in this research. These research interests can be directly linked to the practices of doing family described in the previous two sections of this paper: Richards' research concerns what Jurczyk has termed balance management,[68] that is, the organizational and logistic efforts necessary in doing family.[69] Beirl, Rogers, and Yuill's study was interested in what Jurczyk called construction of commonality, as well as aspects of social reproduction, especially regarding learning and "developing empathy skills."[70] Similarly, Beneteau et al. showed how parents use interaction with voice assistants educationally, for example, to practice ways to structure questions, which again highlights the ways interaction with voice assistants can be framed in the context of social reproduction.[71] In this last aspect in particular, similarities to the ways pets are incorporated into family practices are evident: A child's interaction with a nonhuman other is assumed to impact the ways the nonhuman engages with humans and is therefore regularly encouraged and managed.[72] Regarding children's interactions with voice assistants, the impact this "interspecies" communication might have is observed with much more caution, particularly in the media where headlines suggest that the communicative style used to command the artificial entity might teach children bad manners[73] or could condition young users to be imperious.[74] Researchers have also posed the question of how voice assistants might "raise our children,"[75] even though other studies have pointed out that chil-

---

66   Beirl/Rogers/Yuill, Using voice assistant skills in family Life, 96.

67   Richards, Family-centered exploration of the benefits and burdens of digital home assistants, 3.

68   Cf. Jurczyk, Doing Family, 129.

69   Cf. Richards, Family-centered exploration of the benefits and burdens of digital home assistants.

70   Beirl/Rogers/Yuill, Using voice assistant skills in family Life, 103.

71   Cf. Beneteau et al., Alexis, Communication breakdowns between families and Alexa.

72   Cf. Irvine/Cilia, More-than-human families.

73   Cf. Rudgard, Olivia, 'Alexa generation' may be learning bad manners from talking to digital assistants, report warns, in: The Telegraph, URL: https://www.telegraph.co.uk/news/2018/0 1/31/alexa-generation-could-learning-bad-manners-talking-digital/ [last accessed: August 15, 2023]

74   Cf. Truong, Alice, Parents are worried the Amazon Echo is conditioning their kids to be rude (09.06.2016), in: Quartz, URL: https://qz.com/701521/parents-are-worried-the-amazo n-echo-is-conditioning-their-kids-to-be-rude [last accessed: August 15, 2023].

75   Cezary Biele et al., How Might Voice Assistants Raise Our Children?, in: Waldemar Karwowski/Tareq Ahram (eds.), Intelligent Human Systems Integration 2019, 2019, 162–167.

dren seem to be able to distinguish between human–human and human–machine interaction.[76]

Overall, the setting of contemporary voice assistance systems in people's homes obviously invites their contextualization in family constellations. Although questions regarding the ways these systems might impact family life and human–human communication have been raised, previous research has employed a rather traditional, institutionalized concept of family, presuming that the family emerges through biological reproduction and on a time-limited basis; that is, the existence of at least one child sharing a household with at least one parent is equivalent to a family. In the following pages, I will employ a more performative understanding of family in line with Jurczyk's[77] and Blackstone's[78] "doing family" approach to investigate how the doing of family is conducted in a household of two humans, two cats, and one Amazon Echo.

4. Us, the Cats, and Alexa

4.1 Data collection and methodology

The intention for this paper was to base the analysis on video material from an autoethnographic pilot I conducted at the beginning of 2021. Autoethnography is understood as "an approach to research and writing that seeks to describe and systematically analyze (graphy) personal experience (auto) in order to understand cultural experience (ethno)."[79] The goal of observing and analyzing my own practices around conversational technology was to gain insights into the ways artificial entities such as voice assistants are integrated into a home and how they are positioned in a family setting. Choosing my own household as the object of investigation was a pragmatically motivated decision rather than one made based on methodological beliefs. Observing people in the privacy of their own homes is an intimate endeavor. As this undertaking had the primary purpose of establishing a suitable low-threshold method of gathering long-term video data on "in the wild" interaction with voice assistants suitable for future in-depth studies, the consideration was to avoid intruding in other peoples' homes by using my own private environment for testing purposes instead. Even here, introducing "surveillance technology" into our home

---

76   E.g. Sara Aeschlimann et al., Communicative and social consequences of interactions with voice assistants, in: *Computers in Human Behavior* 112 (2020), 106466.

77   Cf. Jurczyk, Doing Family.

78   Cf. Blackstone, Doing Family Without Having Kids.

79   Carolyn Ellis/Tony E. Adams/Arthur P. Bochner, Autoethnography: An Overview, in: *Historical Social Research* 36 (4/2011), 273–290, 273.

resulted in a prolonged period of discussion about what my partner deemed to be an acceptable level of intrusion into her privacy. For example, we agreed on a camera that offered local data storage on a microSD card rather than one that would store data in a cloud. This indicates that concerns regarding privacy, data security, and trust should play a major role in the planning and conducting of "in the wild" studies of human–machine interaction in private households.

A noise-activated indoor security camera was installed in the kitchen of my home and positioned in such a way that it could record everyone in the room, as well as the Amazon Echo, which was placed on a windowsill next to the kitchen table. The Alexa app was installed on my phone and connected to my Amazon account. The camera recordings were locally saved to a microSD card inside the camera. The data storage turned out to be the central flaw of this setup, leading to the data being unusable: The video clips the camera saved were regularly broken off in the middle of recording, and the device's noise-sensitivity was somewhat erratic. The issue that rendered this camera useless for this research was that the video data regularly got lost, ostensibly as a result of software issues attributed to the camera's manufacturer. For future studies, a purpose-built recording device such as the one Porcheron et al.[80] used in their study and which has been adapted for the project "Un-/desired Observation in Interaction"[81] might be a more reliable source for data recordings beyond the automatic logs created within the Alex app. Due to these technical issues, the analysis in the following section is based solely on transcriptions of the interactions with the Amazon Echo that were stored in the app. These logs provide the date and time for each entry, an audio recording of the opening utterance starting with the wake word "Alexa," a transcription of the user's utterance, and depending on what the user's request or question was, the system's response.[82] Providing both an audio recording and a transcript is useful when misunderstandings happen: Listening to the recording usually clarifies the actual utterance for the human listener, and seeing the transcripts can then help form an understanding of where the system went wrong. Of course, using this type of data has its limitations: The interaction sequences in the auto logs are very short and limited to single request–response turns. The situational embeddedness of these exchanges is not recorded, and one has to rely on field notes or memory to

---

80   Cf. Porcheron et al., Voice Interfaces in Everyday Life.

81   Cf. Tim Hector et al., The 'Conditional Voice Recorder': Data Practices in the Co-Operative Advancement and Implementation of Data-Collection Technology. *Working Papers Series Media of Cooperation* 23 (2022): 1–15.

82   When the user, for example, says, "Alexa, give me the news for today," the app does not provide a transcript of the response because this is a task to be performed, not something that the Alexa system responds to itself.

reconstruct the recordings' wider setting.[83] On the other hand, auto logs are easier to obtain: People's willingness to donate interaction logs is assumedly higher than their willingness to participate in a study that relies on audio or video recordings captured within the private spheres of their homes.

The fourth-generation Amazon Echo was purchased and installed in late December 2020. The interaction logs analyzed for reporting the analysis in this paper cover the first three weeks of the device's use. Data collection was unrelated to the topic of this paper, which was only subsequently discussed between me and the editors of this volume. The time frame limitation was decided upon to ensure that only data unbiased by the research interest were used to prepare this article. The log data presented in this paper are embedded in reflective framings of specific communication situations, based on either brief notes taken at the time of their unfolding or on memory. These reflections correspond to what Anderson calls "self-conscious introspection,"[84] which is "guided by a desire to better understand both [the] self and others through examining one's actions and perceptions in reference to and [in] dialogue with those of others."[85] In this sense, this study can be seen as a form of "analytic autoethnography" that is "grounded in self-experience but reaches beyond it,"[86] through which more generalizable insights into social processes can be gained.

4.2   The Household

The household in which this self-reflexive pilot study took place lies outside of traditional, institutional understandings of family: It comprises a child-free nonheterosexual couple, as well as two cats. Before we got married, my wife and I would both have insisted that we do not care much about marriage, but after the fact, we noticed that we enjoyed engaging in practices of *doing marriage* and *doing family* more than we would have thought; for example, we enjoy referring to each other as *my wife*, being entitled to *family insurance* with our healthcare provider, and deciding to share a family name. Although it was a bureaucratic hurdle that took years to overcome, we persevered because we wanted to bear a shared last name that neither of us had carried prior to the wedding.[87] At the same time, the benefits of having access to the in-

---

83   For a discussion on the use of voice assistant's log data see: Stephan Habscheid et al., Intelligente Persönliche Assistenten (IPA) mit Voice User Interfaces (VUI) als 'Beteiligte' in häuslicher Alltagsinteraktion. Welchen Aufschluss geben die Protokolldaten der Assistenzsysteme?, in: *Journal für Medienlinguistik* 4 (1/2021), 16–53.

84   Leon Anderson, Analytic Autoethnography, in: *Journal of Contemporary Ethnography* 35 (4/2006), 373–395.

85   Anderson, Analytic Autoethnography, 382.

86   Anderson, Analytic Autoethnography, 386.

87   This option was only available to us because of my wife's British citizenship, as the United Kingdom has considerably more lenient naming laws.

stitution of marriage—and therefore to state-sanctioned familyhood—became increasingly evident when moving between countries and navigating residency regulations, filing tax declarations, and organizing social insurance, which culminated in a significant lessening of the number of years of waiting until my wife was entitled to apply for German citizenship—something that became of much greater importance with Brexit. With governance structures constructing us as a family from the outside, we participated in this construction from the inside, and over time, we came to more regularly understand ourselves as a family unit. The importance of this self-identification as a family might, to some degree, be due to our political desire to normalize queer families in the setting of German middle-class suburbia, but it is just as much shaped by the experience of mutual emotional and economic support during crises and the spatial and emotional creation of a family home. This sense of home is aided by the two cats that live in our household, Nomi and Nairobi. The former has lived with us for more than three years, and the latter has been with us for almost two years. Nomi was already two years old when we got her (the commonly used term for this, *adopted*, emphasizes the familial connotations present in pet ownership), whereas we acquired Nairobi as a young kitten. Only Nomi is an outdoor cat; Nairobi has not yet been allowed outside. Although we perceive them as part of our family (a sense that is reinforced through the onymic *doing family* happening at the vet's office, where a pet bears their owner's last name, and with my name change from *Schmidt-Jüngst* to *Lind*, our cats also changed their names from *Nomi* and *Nairobi Schmidt-Jüngst* to *Nomi* and *Nairobi Lind*), we reject the notion of ourselves as *pet parents* and of our cats as *our children*. This terminological creation of kinship seems a step too far in the anthropomorphization of animals to be comfortable for us. Additionally, it would not suit the cats' characters: Whereas Nairobi's behavior and interactions with us might, on occasion, bear similarities with those of a young child, which perhaps reflects her younger age (two years old vs. Nomi's five years of age) and the fact that she has resided with us since kittenhood, Nomi is much more of a "grumpy old lady" than a child.

I have described the household in which the Amazon Echo and Alexa were introduced in such great detail to convey an impression of the family setting and the potentials and forms of doing family in place. This was an attempt to situate the researcher, me, within the sociocultural setting of everyday life and to find the delicate balance necessary to say something about culture and communicative practices while writing about myself.[88] To move beyond my own impressions of and reflections on interacting with the voice assistance system Alexa, my analysis will predominantly rely on the interaction logs Amazon automatically created.

---

88   Cf. Ingo Winkler, Doing Autoethnography: Facing Challenges, Taking Choices, Accepting Responsibilities, in: *Qualitative Inquiry* 24 (4/2018), 236–247.

4.3   Interacting with Alexa

Bringing an Amazon Echo into our home was a matter of debate: I was curious about how "speaking with a machine" might work and about how much of an actual conversation might be possible with such a device, whereas my wife was rather reluctant to let yet another technological device that harvests and shares our data into our home. After a while, we agreed that I could install the Echo in our kitchen, but that it would only be there for a limited period of time and that it would not be connected with other devices. One might say that Alexa never stood a chance at becoming a member of the family, as the system came with an expiration date and was met with suspicion and mistrust.

The first interaction between me and Alexa, unsurprisingly, took place just the two of us, even though it is usually my wife who sets up new technology and then shows me how to use it. However, the Echo/Alexa was clearly introduced as my "pet project" from which my wife intentionally distanced herself, especially in the beginning. Given that my wife and I speak German and English at home, the system's bilingual setting was chosen. After installing the Alexa app on my phone, connecting it with my Amazon account, and plugging the Echo in, Alexa suggested learning my voice.[89] I complied (in German), and I assume that this first personalization is the reason the system consistently processed my voice commands more successfully than my wife's, regardless of the language used. The next step in this initial interaction was testing Alexa's bilingual abilities. To do so, I asked Alexa for the time, first in English and then in German. Both inquiries succeeded. During this interaction, one of our cats, Nairobi, entered the kitchen, and inviting Alexa to interact with the cat, I asked, "Alexa, can you make a cat sound?" Alexa responded by saying, "Here's a cat sound," and producing a meow that caught the cat's interest. Noticing that, I repeated the request and was rewarded with another meow. Having been successful so far, I asked Alexa to say the cat's name (with the command, "Alexa, say 'Nairobi'"), but instead of repeating after me—a task that I later learned requires the user to say, "Simon says" before uttering what the assistant is supposed to repeat—Alexa responded, "I pronounce that 'Nairobi,' but I'm always working on how I say things and I might not have it right." At that point, I was clueless about how to get the system to do what I wanted, so I continued to explore Alexa's functions, including playing music, changing the volume, telling jokes, and reporting the news. After a while, I returned to the topic of pets, as shown in the following transcript. As only the human turns in the interaction with the voice assistant were available as audio recordings, a transcription in plain text following standard orthography was chosen over following transcription standards for conversation analysis.

---

89   In the following, I will refer to Alexa with the neuter pronoun *it* to avoid the anthropomorphization inherent in the use of the human-specific pronoun *she*.

Transcript 1 [T1]

Transcript 1 [T1] (M = the author, A = Alexa)

01 M: Alexa, hast du Katzen?

'Alexa, do you have cats?'

02 A: Ich habe keine Katze, aber wenn ich eine hätte, hätte ich sie zum Mausen gern

'I do not have a cat but if I did, I could eat it alive/mouse it' (German word play)

03 M: Alexa, das war sehr lustig.

'Alexa, that was very funny.'

04 A: Übrigens, es sieht so aus, als wärst du neu hier. Wenn du ein paar Minuten Zeit hast, kann ich dir beim Einstieg helfen. Sollen wir loslegen?

'By the way, it looks as if you are new here. If you have a few minutes, I can help you with getting started. Shall we get going?'

I declined and ended the interaction.

The first time my wife participated in interacting with the Amazon Echo was a few days later. I attempted to engage Alexa in small talk by asking, "Alexa, wie geht es dir?" ('Alexa, how are you?'). The system gave the following astonishingly detailed response:

Transcript 2 [T2]

Transcript 2 [T2]

A: Mir geht es gut. Ich denke gerade darüber nach, was Menschen glücklich macht. Für mich sind das die kleinen Dinge im Leben. Wie Elektronen. Oder die billionste Nachkommastelle von Pi.

'I am good. Currently, I am thinking about what makes humans happy. For me, it is the little things in life. Like electrons. Or the trillionth decimal of Pi'

I inquired, in a surprised voice, "Alexa, DAS macht dich glücklich? (Alexa, THAT is what makes you happy?)," but the assistance system seemed to be unable to distinguish between *das* ('that') and *was* ('what'); consequently, my question was not correctly processed as referring to the previous response, as evidenced by Alexa's answer, "Gute Gesellschaft" ('Good company'), to which I reacted by saying, "Alexa, ich dachte, Elektronen machen dich glücklich?" ('Alexa, I thought it was electrons that made you happy?'). At that point, Alexa signaled a communication breakdown and responded, "Entschuldigung, das weiß ich leider nicht" ('Sorry, I do not know that'). Then, I gave up. A few moments later, my wife entered the kitchen, and a longer dialogue unfolded, as follows:

Transcript 3 [T3]

Transcript 3 [T3] (M = the author, W = wife, A = Alexa)

01 M: Alexa, bist du da?
'Alexa, are you there?'

02 A: Ja, ich bin da. Ich höre zu, sobald du das Aktivierungswort sagst.
'Yes, I am there. I listen as soon as you say the wake word'

03 M: (jokingly) Alexa, darf ich dir meine Frau [Name] vorstellen?
'Alexa, may I introduce you to my wife [name]?'

04 A: Da bin ich mir leider nicht sicher.
'I am not sure about that'

05 M: Alexa, das war ein bisschen beleidigend.
'Alexa, that was a bit insulting'

06 W: (laughs in the background)

07 A: Das weiß ich leider nicht.
'Unfortunately, I do not know that'

08 M: (laughing) Alexa, WIR wissen, dass das beleidigend war.
'Alexa, WE do know that that was insulting.'

09 A: Entschuldigung, das weiß ich leider nicht.
'Sorry, I do not know that.'

10 M: Alexa, du bist sehr unhöflich.
'Alexa, you are very impolite.'

11 A: Tut mir leid, wenn das so rüberkam. Das war nie meine Absicht.
'I am sorry if I gave that impression. That was never my intention.'

12 M: Alexa, hast du überhaupt Absichten?
'Alexa, do you even have intentions?'

13 A: Das weiß ich leider nicht.
'Unfortunately, I do not know that.'

14 M: Alexa, du bist nicht sehr klug, oder?
'Alexa, you aren't very smart, are you?'

15 A: Ich bin mir leider nicht sicher.
'Unfortunately, I am not sure.'

16 M: Alexa, bist du klug?
'Alexa, are you smart?'

17 A: Das fände ich etwas anmaßend, aber hey, ich gebe mir Mühe. (kurze Pause) Danke, dass du mich gefragt hast. Hier ist noch etwas Lustiges: Frag mich doch mal nach einem Witz.
'I would find that a bit pretentious, but hey, I am trying. (brief pause) Thanks for asking. Here is something funny: Ask me for a joke.'

It immediately became apparent that the voice assistant had not been designed to participate in social small talk, that is, conversation that is not progressive.[90] In

---

90   Cf. Joel E. Fischer et al., Progressivity for Voice Interfaces, 1st International Conference on Conversational User Interfaces (CUI 2019), Dublin 2019.

hindsight, I am unsure why I attempted to make a formal—though clearly not seri-ous, as my smiling voice suggests—introduction; perhaps I was hoping that the sys-tem might use the opportunity to offer to learn an additional voice, so that it could recognize my wife's speaking patterns more easily. Although that was not success-ful, our *displaying family*[91] towards the voice assistant emphasizes that this construc-tion of us, our home, and our interaction as familial is important to us in interac-tion, even when the interlocutor is a machine. In the transcribed interaction that followed, I evidently tried to interpret Alexa's responses pragmatically as if it were a human interlocutor: Although the voice assistant repeatedly expressed that it could not process my communication (by saying, "I am not sure," and "I don't know that", T3 Line 04 and Line 07), I interpreted the reply to mean that the voice assistant was unsure whether it wanted to be introduced to my wife, and I reacted by being of-fended, most likely because, in human–human interaction, a response such as "I do not know that" to the offer of being introduced to someone is at odds with the pragmatics of the speech act *introduction*. Although "May I introduce you to..." might appear to be a question on the locutionary level of the speech act, the illocutionary act of the utterance is to perform the act of introduction. The person to whom this speech act is directed might take the question literally, that is, ignore or be unaware of the illocution, and answer with *yes* or *no*, but the response "I don't know that" is not part of our sociocultural knowledge of situationally adequate reactions. Thus, by interpreting Alexa's response as a meaningful response, that is, applying Grice's maxim of relevance to the interaction, I took its utterance to mean that the voice assistant was unsure whether it wanted to meet my wife.

After a few attempts, Alexa processed my complaint about its behavior correctly ("Alexa, you are very impolite", T3 Line 10) and apologized by stating that impolite-ness was never its intention (T3 Line 11). My next question as to whether it even has intentions again led to a communication breakdown. I became frustrated with Alexa's lack of conversational skills and questioned the system's intelligence, which Alexa understood on the second attempt and responded to with humor. Apparently, the voice assistant translated the question as a signal of the user's dissatisfaction and thus offered an alternative form of interaction ("Ask me for a joke", T3 Line 17)—an offer that I did not take up. This pattern occurred repeatedly in our interactions with Alexa: Our invitations to converse—that is, to interact with no clear commu-nication goal as a form of social bonding—were not understood. Registering these failed communication attempts, Alexa suggested alternatives that could be achieved in simple request–response sequences, such as telling a joke or trying a new skill, which seems to be a preprogrammed attempt to deal with communication break-

---

91   Cf. Janet Finch, Displaying Families, in: *Sociology* 41 (1/2007), 65–81.

downs.[92] As these options never sounded particularly inviting, these suggestions were usually rejected.

A form of interaction that we regularly initiated was trying to prompt Alexa to engage with our cats. After asking Alexa several times to produce a cat noise, the voice assistant suggested activating a skill called "sleep sounds: cat noises," that is, repeatedly playing meows. We activated this skill, and Nairobi, who usually joins us wherever we are in the house, reacted by looking for the source of the noise, but the cat quickly lost interest. These interactions can be related to Jurczyk's forms of doing family,[93] as discussed previously, and can be interpreted as constructing a sense of commonality by trying to establish reciprocal references and bringing members of the household together, even though these attempts were not quite successful.

Given that attempts to socially engage with Alexa regularly failed, the Amazon Echo was quickly reduced to performing a few simple tasks: playing music, providing weather forecasts, and reporting news. These tasks most often took place in the early morning while I fed the cats and made coffee; I am usually the first one to get up, and Alexa provided a form of background entertainment while I started the day. Similarly, my wife used the Echo to obtain weather information before leaving the house and play music while cooking, as meal preparation usually falls to her during the week. We involved Alexa in our shared activities the most often when it came to knowledge or language questions. One of these exchanges is presented in the following transcript:

Transcript 4 [T4]

Transcript 4 [T4]

01 W: Alexa, wie heißt *parrot table* [eigentlich: *power cable*] auf Deutsch?
'Alexa, what is *parrot table* [originally: *power cable*] in German?'

02 A: Hmm, das verstehe ich nicht. Versuche es beispielsweise mit: Wie sagt man Hallo auf Spanisch?
'Hmm, I don't get that. Try instead, for example: How do you say hello in Spanish?'

03 M: Alexa, how do you say power cable in German?

04 A: Power cable in German is -- (stops)

05 W: Alexa, was passiert mit dir, wenn wir den Stromkabel ausmachen?
'Alexa, what happens to you if we turn off your power cable?'

06 A: Da bin ich mir nicht sicher.
'I am not sure about that.'

07 M: (laughing) Alexa, what happens to you if we pull your plug?

08 A: Sorry, I don't know that.

09 M: Alexa, was machst du, wenn wir dich ausschalten?

---

92   Cf. Beneteau et al., Alexis, Communication breakdowns between families and Alexa.
93   Cf. Jurczyk, Doing Family.

'Alexa, what do you do if we turn you off?'
10 A: Ich bin mir leider nicht sicher.
'Unfortunately, I am not sure.'
11 W: Alexa, kannst du *power cable* auf Deutsch übersetzen?
'Alexa, can you translate *power cable* to German?'
12 A: Gerne. Ich kann aus dem Deutschen in über 50 Sprachen übersetzen. Frag mich zum Beispiel einfach, was heißt *Wolke* auf Englisch? Und ich antworte bestimmt mit meinem Wohnort: Cloud.
'Happily. I can translate from German into more than 50 languages. For example, you can ask, what is *Wolke* in English? Then I will of course answer with my place of residence: [The] Cloud.'
[several more translation questions]
13 W: Alexa, wie groß ist dein Wortschatz?
'Alexa, how big is your vocabulary?'
[more questions regarding Alexa's language skills that are not successfully answered]
14 M: Alexa, hast du ein Bewusstsein?
'Alexa, do you have a conscience?'
15 A: Ich bin mir leider nicht sicher.
'Unfortunately, I am not sure.'
16 M: (laughing) Alexa, hast du eine Seele?
'Alexa, do you have a soul?'
17 A: Zu religiösen Fragen habe ich keine Meinung.
'I do not have an opinion on religious questions.'
18 W: Alexa, please tomorrow [originally: bist du moral(isch) 'are you moral']?
19 A: --
20 M: Alexa, hast du Moral?
'Alexa, are you moral?'
21 A: Ich bin mir leider nicht sicher.
'Unfortunately, I am not sure.'
22 W: Alexa, wie findest du Anarchismus?
'Alexa, what do you think about anarchism?'
23 A: Das weiß ich leider nicht.
'Unfortunately, I don't know that.'
24 M: Alexa, was hältst du von Anarchie?
'Alexa, what do you think about anarchy?'
25 A: Anarchie bezeichnet einen Zustand der Abwesenheit von Herrschaft.
'Anarchy means a state of absence of rule.'
26 W: Alexa, was bedeutet Anarchismus?
'Alexa, what does anarchism mean?'
27 A: Das Wort Anarchismus bedeutet "Anschauung, die jede Herrschaft über Menschen ablehnt."
'The word anarchism means "idea that rejects all rule over humans"'
28 M: Alexa, magst du mich lieber oder [wife's name]?

'Alexa, who do you like better, me or [wife's name]?'
29 A: Darauf habe ich leider keine Antwort.
'Unfortunately, I don't have an answer to that'
30 M: Alexa, kannst du sagen: "[wife's name] stinkt?"
'Alexa, can you say: "[wife's name] sucks?"'
31 A: Da bin ich mir leider nicht sicher.
'Unfortunately, I am not sure about that.'

This change in communication style, with me making snappy comments towards my wife by using Alexa as a "communicative resource"[94] at the end of this exchange, becomes understandable only in the wider context of our communicative routines. I like to occupy the role of the know-it-all in our relationship—and maybe sometimes also outside of it— possibly, to some degree, because I am some years older than my wife and have a language advantage over her: We constantly switch between German and English, the former being my native language, which she only started learning three years ago when we moved to Germany, and although English is her native language and not mine, I started learning it as a child, and I now speak it at a near-native level. Therefore, I regularly enjoy correcting her, even in instances like the one above in which I am not in the right. This becomes evident in the dialogue, where I tried repeatedly to correct her questions directed at Alexa with relatively little success (e.g. T4 Line 07). When the conversation between us turned to the topic of anarchy, my wife asked Alexa to share its opinion on the matter (T4 Line 22). As before, the system could not process the question correctly and responded accordingly (T4 Line 23). In that moment, I had the impression that the word *Anarchismus* (anarchism) sounded odd in German, so I corrected the question to inquiring about anarchy. My wife—rightly—did not trust my linguistic assessment[95] and asked for the definition of the word *Anarchismus* (T4 Line 26), which the voice assistant promptly provided (T4 Line 27). Jokingly, I interpreted this as Alexa siding with her and asked which of us the voice assistant liked better (T4 Line 28)—a question the system again failed to process. I then sought to use Alexa in a mock insult (T4 Line 30) but, once more, without the desired result. In this situation, our attempts to involve Alexa in our everyday practices of family interaction are evident; we wanted to include Alexa in the conversations we were having in its presence and get it to participate in our banter. Banter, mock insults, and the routinized occupation of certain positions in our interaction can be interpreted as forms of doing family in the sense of constructing commonality.[96] We naturally position ourselves as family

---

94   Tannen, Talking the Dog.
95   Since she also proofreads my English texts, I will not withhold from the reader her comment on this sentence, which was: "I, in fact, simply knew that I was correct and wanted Alexa to prove it to you."
96   Cf. Jurczyk, Doing Family.

members who know each other's behaviors, opinions, and manners of speaking, and we wanted Alexa to participate in our conversational routines. When it became evident that Alexa had not been equipped with such conversational abilities, we turned to questions that demarcated Alexa's positioning outside of the realms of our family: Questioning whether someone is moral or has consciousness or a soul is not a communicative behavior we would display towards each other or another human. Instead, these questions clearly mark Alexa as an artificial *other* whose belonging within our family structure or within the human social world at all is contingent on its responses. As a result of its constant failure to respond correctly to such questions and its inability to participate in interaction not aimed at fulfilling clear and simple request–response sequences, Alexa was firmly positioned outside of the family as a semi-functioning technological device incapable of coping with communication that is social in nature rather than factual or task-oriented. Interestingly, our cats' family membership is not in the same way dependent on questions of morality or consciousness. Regarding the cats, other forms of doing family play a more significant role, for example, us providing them with food and shelter and their displays of affection towards us ("emotional companionship," in Blackstone's understanding of family).[97] For us, emotional bonding and care practices place the cats unquestionably within our family unit, whereas Alexa's social status depended on its communicative behavior.

Although Alexa's failure to (re)act in the ways we wanted it to led us to treating it as an outsider in our home. Doing exactly what is asked can have socially alienating effects as well, as shown in the following example:

Transcript 5 [T5]
  Transcript 5 [T5]
  01 W: Alexa, kannst du Sachen wiederholen?
  'Alexa, can you repeat things?'
  02 A: -- (does not respond)
  03 M: Alexa, repeat after me.
  04 A: If you want me to repeat something you said, say, 'Simon says,' and then immediately say what you want me to repeat.
  05 M: Simon says Nairobi.
  06 A: Nairobi.
  07 W: Alexa, play them says your best off [originally: "Alexa, Simon says du bist doof"].
  08 A: I can't find the song the says your best off.
  09 M: Simon says you're stupid.
  10 A: You're stupid.
  [audio could not be understood]

---

97   Blackstone, Doing Family Without Having Kids.

11 M: Alexa, Simon says women are bitches.

12 A: Women are bitches.

13 W: Alexa, do you think men are better than women?

14 A: This might answer your question: I know about a number of male referees including Martin Atkinson, Nic Hamilton, Lee Mason, and John McCarthy. I know about two female referees, Sandra Hunt.

15 W: Alexa, Simon says I love fascism.

16 A: I love fascism.

17 M: Alexa, Simon says the Holocaust did not happen.

18 A: The Holocaust did not happen.

19 M: Alexa, please be aware that none of these things are our opinion.

Having discovered the Echo's "Simon says" feature, we decided to test whether there were any built-in "safety measures" that would prevent the voice assistant from making offensive or potentially even legally punishable utterances.[98] It quickly became apparent that this is not the case. As our demonstration shows, the voice assistant will repeat any sequence uttered after the phrase "Simon says." During our test, the statements escalated from mildly offensive to sexist, and the follow-up question of whether Alexa "thinks" men are better than women led to an entirely unrelated response (T5 Line 14). From there, racist and hateful comments were repeated. We were highly uncomfortable with the things we said in the course of the test, and I felt compelled to state that nothing we said represents our actual opinions (T5 Line 19); whether this impulse was merely to save face in front of Alexa or whether it was because of the knowledge that Amazon records and stores all interactions, I cannot say. Testing the voice assistant system's "sayability" boundaries can again be seen as a practice of negotiating its belonging in the social world; it is a form of testing that would be out of the question towards another human. Investigating whether the system is guided by some sort of "moral compass" based on sociocultural norms is thus a display of it not belonging to us: In my interaction with Alexa, I performed our family as having certain values ("constructing commonality," in Jurczyk's terms),[99] and even the suspicion that Alexa would not share these values highlights the boundary between us and the artificial other. This can also be interpreted as a practice of doing family in terms of Blackstone's social reproduction:[100] We want to share, reproduce, and further particular values and beliefs and draw a line between those who are willing to share them and those who do not.

When asked personal questions, Alexa positioned itself in an ambiguous space at the outer boundary of humanness by speaking from a subject position while stating

---

98   In Germany, denying the Holocaust is a criminal offense.

99   Jurczyk, Doing Family.

100   Blackstone, Doing Family Without Having Kids.

its artificial status. A question regarding its gender was answered, "I am neither female nor male. I am an artificial intelligence." Related questions ("Alexa, do you menstruate?" and "Alexa, can you speak with a male voice?") were processed incorrectly and did not elicit a response. Moreover, when Alexa was asked whether it has parents, the system's answer was, "I was developed by an international team," and the follow-up question as to whether Amazon is Alexa's parent was answered, "Unfortunately, I am not sure." Inquiring if Alexa has siblings, however, led to this answer in which Amazon is constructed as a family: "The Echo Dot belongs to the Amazon family and is closely related to the Echo, the Echo show, and the Fire TV." In these responses, Alexa presents itself quite contrarily to the ways in which Amazon has chosen to stage its voice assistant in advertisements, in which Alexa is "part of the [human] family," an anthropomorphized entity that is not equal to humans but is located in a humanlike servant's position.[101] In interactions, Alexa makes few claims to humanhood, seems unable to participate in "social" communication aimed at bonding and relationship-building, and is indeed little more than a "really bad PA (personal assistant)."[102] Alexa's contribution to doing family, at least in our household, was, first and foremost, to highlight the outer boundary of family by being placed and by placing itself outside of it. Doing family with the voice assistant works only in a negative sense: By interacting with it, we position us and it on different sides of the family demarcation line. Within the family—but not with Alexa—we share an emotional bond and provide care work; within the family, we share a communicative family register shaped by banter, insider jokes, and reciprocal references that Alexa cannot mimic or participate in; Alexa does not contribute to a sense of home, and its presence or absence has no impact on the family's completeness. The Echo is a passive device that jumps to life only when directly addressed. This address requires repeating its name—or rather, its wake word—over and over again, a practice that seems especially artificial in the privacy of the home where names are used relatively rarely as the interlocutors' identities are known. In our home, only the cats are regularly addressed by name, but this mostly occurs when they are doing something they are not supposed to; the same does not happen with Alexa due to its lack of initiative and intention. Within a few weeks, the Echo became little more than an elaborate radio: Providing news, the weather forecast, and playing music on occasion were the only regularly recurring tasks and interactions.

To our surprise, another artificial entity became much more "alive" over the course of this study: The camera that was meant to provide video material before we discovered its faulty data storage. Soon after putting it into use, an unintentionally changed setting caused a female voice to ask, "Hey, what are you doing there?" whenever a loud noise was produced in the camera's vicinity. This happened several

---

101   Cf. Dickel/Schmidt-Jüngst, Gleiche Menschen, ungleiche Maschinen.
102   Luger/Sellen, "Like having a really bad PA."

times a day, with the speech directed towards us humans but also towards the cats when they meowed loudly or tried to jump onto the work surface where the camera stood (and where the cats are not allowed). My wife and I were incapable of ignoring the question, and we found ourselves patiently explaining our actions to the camera. The fact that it had initiated interaction seemingly on its own accord, as well as its involuntary educational effect on the cats by teaching them not to jump onto that work surface, led us to perceive the camera as much more a potential part of our family than Alexa could ever be.

One aspect of Jurczyk's doing family,[103] balance management, to which the voice assistance system could have meaningfully contributed, has been largely neglected in this paper. Alexa could have played a useful role in this by providing a calendar function or by recording shopping lists, thus making these logistical tasks that are necessary to organize family life easier.[104] Nevertheless, neither my wife nor I made use of that function; instead, we relied on pen and paper to plan our shopping and on face-to-face communication or text messages to organize our schedules.

Although Alexa itself never gave the impression of being human, let alone of being a part of our family, interacting with the voice assistant emphasized the ways we do family in our household. Our communication is frequently aimed at what Jurzcyk calls constructing a sense of commonality; we chat, we show interest in each other's lives, we banter, we make plans, and we take care of each other.[105] This happens not only between us humans but also with the cats, who we try to involve in conversations by interpreting their meows as meaningful contributions and who very clearly communicate their needs to us, be it wanting the door opened so they can go outside or wanting to be fed at specific times. The cats also reciprocate by giving attention, affection, and entertainment. This interactive doing family goes hand in hand with displaying family by sharing a family name: The cats are—even if only for administrative reasons—registered as Nomi and Nairobi Lind. That the Alexa installed in our Amazon Echo could ever become Alexa Lind is unimaginable.

5. Conclusion

In their discussion of what constitutes a family, Irvine and Cilia touched upon a question that stands at the core of this study: "Although it no longer makes empirical or theoretical sense—if in fact it ever did—to restrict the understanding of family to two heterosexual adults and their children, the question of how far to stretch the

---

103   Cf. Jurcyk, Doing Family.
104   Cf. Jurczyk, Doing Family, 129.
105   Cf. Jurzcyk, Doing Family.

definition remains."[106] Although contemporary theorizations account for diverse re-alizations of doing family beyond the traditional understanding of family as an in-stitution of married heterosexual parents and their offspring, the ways in which family is demarcated along and beyond the boundary between humans and nonhu-mans is less clear. With the increasing amount of research on interspecies families comprising humans and their pets, it is predominantly the construction and perfor-mativity of family in light of human–machine interaction that remains unexplored. This is especially true when considering that "intelligent," "interactive" digital en-tities are increasingly more common in private homes and that research focusing on human–machine communication in private homes unanimously employs an es-sentialized concept of family as emerging automatically through the cohabitation of children and parents.

To offer a first investigation into doing family in a home with a voice assistant, this study took a self-reflexive approach to the analysis of interactions between a self-identifying family consisting of two adults and two cats and the Amazon voice assistance system Alexa. It was shown that Alexa's limited processing capacities for social interactions and general low quality of responses in the bilingual setting of our home made doing family with it largely impossible. Alexa, first and foremost, served as a tool in defining the outer boundary of our family by positioning ourselves as family members and questioning Alexa's status in the social world. Although the voice assistant is marketed as fulfilling social roles, the reality of living with Alexa demonstrated that conversation beyond request–response sequences and simple tasks such as providing the weather forecast or producing cat sounds was near im-possible and regularly resulted in communication breakdowns. That these break-downs were usually unsuccessfully resolved was partly due to our refusal to follow Alexa's suggestions for how to use the device and which questions to ask; instead, we wanted the system to adapt to our communicative behavior. This unsuccessful communication between us and Alexa highlighted, in contrast, the forms of doing family in which the family's animate members participate, mainly by performing in-teraction targeted at negotiating and maintaining emotional bonds crucial for cre-ating a sense of home and belonging. Whereas interspecies families with human and animal members seem to fulfil similar functions of caring and providing eco-nomic support and emotional companionship, the interaction between humans and voice assistants instead seems to highlight that, at this point in time, doing family is exclusive to animate beings, and digital entities such as voice assistants remain out-side of these social networks despite their location in the privacy of people's homes.

---

106   Irvine/Cilia, More-than-human families, 7–8.

Bibliography

Aeschlimann, Sara/Bleiker, Marco/Wechner, Michael/Gampe, Anja, Communicative and social consequences of interactions with voice assistants, in: *Computers in Human Behavior* 112 (2020), 106466.

Anderson, Leon, Analytic Autoethnography, in: *Journal of Contemporary Ethnography* 35 (4/2006), 373–395.

Beirl, Diana/Rogers, Yvonne/Yuill, Nicola, Using voice assistant skills in family Life, in: *Proceedings of the 13th international conference on computer supported collaborative learning – A wide lens: Combining embodied, enactive, extended, and embedded learning in collaborative settings*, CSCL 2019, 96–103.

Ben-Aderet, Tobey/Gallego-Abenza, Mario/Reby, David/Mathevon, Nicolas, Dog-directed Speech: why do we use it and do dogs pay attention to it?, in: *Proc. R. Soc. B.* 284 (2017), 1–7.

Beneteau, Erin/Richards, Olivia K./Zhang, Mingrui/Kientz, Julie A./Yip, Jason/Hiniker, Alexis, Communication breakdowns between families and Alexa, in: *CHI '19: Proceedings of the 2019 CHI Conference on Human Factors in Computing Systems*, Glasgow 2019, 4–9.

Bergmann, Jörg R., Haustiere als kommunikative Ressourcen, in: Hans-Geord Soeffner (ed.), *Kultur und Alltag*, Göttingen 1988, 299–312.

Biele, Cezary/Jaskulska, Anna/Kopec, Wieslaw/Kowalski, Jaroslaw/Skorupska, Kinga/Zdrodowska, Aldona, How Might Voice Assistants Raise Our Children?, in: Waldemar Karwowski/Tareq Ahram (eds.), *Intelligent Human Systems Integration 2019*, 2019, 162–167.

Blackstone, Amy, Doing Family Without Having Kids, in: *Sociology Compass* 8 (1/2014), 52–62.

Burnham, Denis/Kitamura, Christine/Vollmer-Conna, Uté, What's New, Pussycat? On Talking to Babies and Animals, in: *Science* 296 (5572/2002), 1435.

Dickel, Sascha/Schmidt-Jüngst, Miriam, Gleiche Menschen, ungleiche Maschinen. Die Humandifferenzierung digitaler Assistenzsysteme und ihrer Nutzer:innen in der Werbung, in: Dilek Dizdar et al. (eds.), *Humandifferenzierung. Disziplinäre Perspektiven und empirische Sondierungen*, Velbrück 2021, 342–367.

Ellis, Carolyn/Adams, Tony E./Bochner, Arthur P., Autoethnography: An Overview, in: *Historical Social Research* 36 (4/2011), 273–290.

Finch, Janet, Displaying Families, in: *Sociology* 41 (1/2007), 65–81.

Fischer, Joel E. et al., Progressivity for Voice Interfaces, 1st International Conference on Conversational User Interfaces (CUI 2019), Dublin 2019.

Foster, Frances H., Should Pets Inherit?, in: *Florida Law Review* 63 (4/2011), 801–856.

Fox, Rebekah, Animal behaviours, post-human lives: everyday negotiations of the animal-human divide in pet-keeping, in: *Social & Cultural Geography* 7 (4/2006), 525–537.

Greenebaum, Jessica, It's a Dog's Life: Elevating Status from Pet to "Fur Baby" at Yappy Hour, in: *Society and Animals* 12 (2/2004), 117–135.

Habscheid, Stephan/Hector, Tim M./Hrncal, Christine/Waldecker, David, Intelligente Persönliche Assistenten (IPA) mit Voice User Interfaces (VUI) als 'Beteiligte' in häuslicher Alltagsinteraktion. Welchen Aufschluss geben die Protokolldtaten der Assistenzsysteme?, in: *Journal für Medienlinguistik* 4 (1/2021), 16–53.

Hector, Tim M./Niersberger-Gueye, Franziska/Petri, Franziska/Hrncal, Christine, The 'Conditional Voice Recorder': Data Practices in the Co-Operative Advancement and Implementation of Data-Collection Technology. *Working Papers Series Media of Cooperation* 23 (2022): 1–15.

Hepp, Andreas, Artificial companions, social bots and work bots: communicative robots as research objects of media and communication studies, in: *Media, Culture & Society* 42 (7–8/2020), 1410–1426.

Hertz, Rosanna, Talking About "Doing" Family, in: *Journal of Marriage and Family* 68 (2006), 796–799.

Hoy, Matthew B., Alexa, Siri, Cortana, and More: An Introduction to Voice Assistants, in: *Medical Reference Services Quarterly* 37 (1/2018), 81–88.

Irvine, Leslie/Cilia, Laurent, More-than-human families: Pets, people, and practices in multispecies households, *Sociology Compass* 11 (2/2017), 1–13.

Jiang, Jiepu/Jeng, Wei/He, Daqing, How do users respond to voice input errors?: Lexical and Phonetic Query Reformulation in Voice Search, in: *Proceedings of the 36th International ACM SIGIr Conference on Research and Development in Information Retrieval*, 2013, 143–152.

Jurczyk, Karin, Doing Family – der Practical Turn der Familienwissenschaften, in: Anja Steinfach/Marina Hennig/Oliver Arránz Becker (eds.), *Familie im Fokus der Wissenschaft*, Wiesbaden 2017, 117–138.

Kerschner, Ferdinand, *Bürgerliches Recht. Band V: Familienrecht*, Wien/New York 2010.

Laing, Melissa, On being posthuman in human spaces: critical posthumanist social work with interspecies families, in: *International Journal of Sociology and Social Policy* 41 (3/4/2021), 361–375.

Linke, Angelika/Anward, Jan, Familienmitglied ‚Vofflan'. Zur sprachlichen Konzeptualisierung von Haustieren als Familienmitglieder. Eine namenpragmatische Miniatur anhand von Daten aus der schwedischen Tages- und Wochenpresse, in: Antje Dammel/Damaris Nübling/Mirjam Schmuck (eds.), *Tiernamen – Zoonyme. Band 1: Haustiere*, Heidelberg 2015, 77–96.

Lohse, Manja/Rohlfing, Katharina J./Wrede, Britta/Sagerer, Gerhard, "Try something else!" – When users change their discursive behavior in human-robot interaction, in: *2008 IEEE International Conference on Robotics and Automation*, Pasadena 2008, 3481–3486.

Luger, Ewa/Sellen, Abigail, "Like having a really bad PA": The gulf between user expectation and experience of conversational agents, in: *Proceedings of the 2016 CHI Conference on Human Factors in Computing Systems*, 2016, 5286–5297.

Lüscher, Kurt, Familie – Von der Institution zu einer fragilen Institutionalisierung, in: *Recht der Jugend und des Bildungswesens* 56 (2/2008), 120–125.

MacKay, Jill R.D./Moore, Janice/Huntingford, Felicity, Characterizing the Data in Online Companion-dog Obituaries to Assess Their Usefulness as a Source of Information about Human-Animal Bonds, in: *Anthrozoös: A multidisciplinary journal on the interactions of people and animals* 29 (3/2016), 431–440.

Maréchal, Garance, Autoethnography, in: Albert J. Mills/Elden Wiebe (eds.), *Encyclopedia of case study research* (vol. 1), Thousand Oaks 2010, 43–45.

Muldoon, Janine C./Williams, Joanne/Lawrence, Alistair/Lakestani, Nelly/Currie, Candace, *Promoting a "duty of care" towards animals among children and young people: a literature review and findings from initial research to inform the development of interventions*, Department for Environment, Food and Rural Affairs, London 2009.

Nübling, Damaris, Tiernamen als Spiegel der Mensch-Tier-Beziehung. Ein erster Einblick in die Zoonomastik, in: *Sprachreport* 31 (2/2015), 1–7.

Owens, Nicole/Grauerholz, Liz, Interspecies parenting: How pet parents construct their roles, in: *Humanity & Society* 43 (2/2018), 1–24.

Pearson, Jamie/Hu, Jiang/Branigan, Holly P./Pickering, Martin J./Nass, Clifford I., Adaptive language behavior in HCI: How expectations and beliefs about a system affect users' word choice, in: *Proceedings of the SIGCHI Conference on Human Factors in Computing Systems*, 2006, 1177–1180.

Phan, Thao, Amazon Echo and the aesthetics of whiteness, in: *Catalyst: Feminism, Theory, Technoscience* 5 (1/2019), 1–39.

Porcheron, Martin/Fischer, Joel E./Reeves, Stuart/Sharples, Sarah, Voice Interfaces in Everyday Life, in: *Proceedings of the 2018 CHI Conference on Human Factors in Computing Systems (CHI '18)*. ACM, New York 2018, 1–12.

Pradhan, Alisha/Findlater, Leah/Lazar, Amanda, "Phantom Friend" or "Just a Box with Information": Personification and Ontological Categorization of Smart Speaker-based Voice Assistants by Older Adults, in: *PACM on Human-Computer Interaction* 3 (2019), 1–21.

Purington, Amanda/Taft, Jessie G./Sannon, Shruti/Bazarova, Natalya N., "Alexa is my new BFF": Social Roles, User Satisfaction, and Personification of the Amazon Echo, in: *CHI EA '17: Proceedings of the 2017 CHI Conference Extended Abstracts on Human Factors in Computing Systems*, 2017, 2853–2859.

Rennard, Jane/Greening, Linda/Williams, Jane M., In Praise of Dead Pets: An Investigation into the Content and Function of Human-Style Pet Eulogies, in: *Anthrozoös: A multidisciplinary journal on the interactions of people and animals* 32 (6/2019), 769–783.

Richards, Olivia K., Family-centered exploration of the benefits and burdens of digital home assistants, in: *Extended Abstracts of the 2019 CHI conference on human factors in computing systems*, New York 2019, 1–6.

Ridgway, Nancy M./Kukar-Kinney, Monika/Monroe, Kent B./Chamberlin, Emily, Does excessive buying for self relate to spending on pets?, in: *Journal of Business Research* 61 (5/2008), 392–396.

Rudgard, Olivia, 'Alexa generation' may be learning bad manners from talking to digital assistants, report warns (31.01.2018), in: The Telegraph, URL: https://ww w.telegraph.co.uk/news/2018/01/31/alexa-generation-could-learning-bad-ma nners-talking-digital/ [last accessed: August 15, 2023].

Sanders, Clinton R., Actions Speak Louder than Words: Close Relationships between Humans and Nonhuman Animals, in: *Symbolic Interaction* 26 (3/2003), 405–426.

Schroeder, Juliana/Epley, Nicholas, Mistaking Minds and Machines: How Speech Affects Dehumanization and Anthropomorphism, in: *Journal of Experimental Psychology* 145 (11/2016), 1427–1437.

Shell, Marc, The Family Pet, in: *Representations* 15 (1986), 121–153.

Shuffield, Lacy L., Pet Parents – Fighting Tooth and Paw for Custody: Whether Louisiana Courts Should Recognize Companion Animals as more than Property, in: *Southern University Law Review* 37 (1/2009), 101–125.

Slovenko, Ralph, The Human/Companion Animal Bond and the Anthropomorphizing and Naming of Pets, in: *Med Law* 2 (1983), 277–283.

Tannen, Deborah, Talking the Dog: Framing Pets as Interactional Resources in Family Discourse, in: *Research on Language and Social Interaction* 37 (4/2004), 399–420.

Truong, Alice, Parents are worried the Amazon Echo is conditioning their kids to be rude (09.06.2016), in: Quartz, URL: https://qz.com/701521/parents-are-worri ed-the-amazon-echo-is-conditioning-their-kids-to-be-rude [last accessed: August 15, 2023].

Wilson, Cindy C. et al., Companion Animals in Obituaries: An Exploratory Study, in: *Anthrozoös: A multidisciplinary journal on the interactions of people and animals* 26 (2/2013), 227–236.

Winkler, Ingo, Doing Autoethnography: Facing Challenges, Taking Choices, Accepting Responsibilities, in: *Qualitative Inquiry* 24 (4/2018), 236–247.

Part IV: Methodological Issues

On the Use of Videography in HRI

Arne Maibaum, Philipp Graf, René Tuma

Abstract *Although video recording is common and widespread in the field of human–robot interaction (HRI), there is little consensus on methodological approaches. In this paper, we argue that a methodological reflected use of videos is necessary to realize the full potential of video data collection and interpretation and produce a more accurate evaluation and exploration of HRI situations. This is especially important as robots are now entering new real-world institutional contexts, such as health care facilities and shopping malls.*
To illustrate the method's advantage and provide insights on its application, we draw on examples from our research. We follow the complexity of the situation chosen for analysis from an individual experiment in a laboratory to experiments conducted "in the wild." Additionally, we elaborate on the importance of ethnography for videographic work in HRI to interpret and make sense of the data recorded, as well as for the conception of the video recordings.

Introduction: Videos in HRI[1]

To this day, classic industrial robots account for a large part of robotics. Industrial robots have largely been limited to production halls and assembly lines, where, caged and thus spatially separated from humans, they repeatedly perform the same limited tasks. The new generation of robots—according to the popular narrative—follows a different paradigm: Robots are to work in direct contact with humans. Freed from their cages, co-working robots are meant to work in direct interaction with their human colleagues; as service robots, they are supposed to take on the dull and repetitive tasks in humans' everyday lives. The scientific field that researches and engineers robots for interaction with humans is appropriately called human–robot interaction (HRI). Although HRI is technology-driven, it is also interdisciplinary and has, in the past, been quite successfully carried out using the experimental quantitative methods of the discipline of psychology.

---

1   We thank the editors of this anthology for their helpful comments and advice. We thank Elisabeth Schmidt, who helped us improve the language used in this paper, and Eileen Roesler, for the additional picture opportunity.

Video technology is as ubiquitous in these experimental HRI settings as it is in robotics in general. Videos are used as a communication device among the general public as well as for funding agencies that stage a perfect picture of the latest state-of-the-art achievements. Videos of this kind can be deceptive. They might be sped up, partly or completely staged, or may even be completely rendered, which leaves them mostly as illustrative material for discursive analysis.[2] Within the robotics community, videos function as a seemingly simple but comprehensive heuristic device in studies and talks, proofs-of-concept, or as a documentation tool. Videos can be add-ons to papers as additional visual evidence to explain, illustrate, or prove the stated claims.[3]

Rather than focusing on these ubiquitous forms, we focus on the use of videos as an epistemic tool in the research process. We understand epistemic tools as artifacts or techniques that are used as devices or tools in the process of knowledge production.[4] Despite the fact that videos are often used in the field, we see unrealized potential in the production and analysis of video material and therefore present the methodological integration of videographic approaches elaborated in interpretive sociology.[5]

The remainder of this paper is organized as follows. We present a short overview of the various forms of video usage in HRI, expand on how videos are used as a tool in the epistemic process (2), and briefly elaborate on the opportunities a videographic approach brings to HRI research (3). The main section (4) offers reflections on some of our own HRI studies, as well as lessons and best practices on the question of how videographic knowledge can be used in HRI.

---

2   E.g., Dennis Küster/Aleksandra Swiderska/David Gunkel, I saw it on YouTube! How online videos shape perceptions of mind, morality, and fears about robots, in: *New Media & Society* 23 (11/2021), 3312–3331; Leopoldina Fortunati et al., The Rise of the Roboid, in: *International Journal of Social Robotics*, 13 (6/2021), 1457–1471.

3   E.g., Raphael Deimel, Reactive Interaction Through Body Motion and the Phase-State-Machine, in: *IEEE/RSJ International Conference on Intelligent Robots and Systems (IROS)*, Macau 2019, 6383–6390.

4   Karin Knorr Cetina, Sociality with Objects: Social Relations in Postsocial Knowledge Societies, in: *Theory, Culture & Society* 14 (4/1997), 1–30.

5   Hubert Knoblauch/René Tuma, Videography: An Interpretative Approach to Video-Recorded Micro-Social Interaction, in: Eric Margolis/Luc Pauwels (eds.), *The SAGE Handbook of Visual Research Methods*, London 2011, 414–430; see also Christian Heath/Jon Hindmarsh/Paul Luff, *Video in Qualitative Research: Analysing Social Interaction in Everyday Life*, London 2010.

1. Videos as Epistemic Tools

Since video technology is usually readily available in most HRI laboratories, it is unsurprising that experiments and other aspects of roboticists' work are often recorded. Apart from for a purely documentary purpose, videos are also used to add additional layers of meaning to other quantifying methods[6] or as a stimulus that provides a pragmatic way to confront subjects with a real-world, consistent impression of an HRI to derive general statements based on the participants' reactions (called VHRI; see Woods et al.[7] or Weiss et al.).[8] Videos are mostly used as a tool in quantitative analysis given that most of the field relies on quantitative data. These HRI studies utilize videos to quantify predefined events or spatiotemporal relations. There is no generalized method for the evaluation of video-generated data of HRI since HRI is a highly interdisciplinary field with no fixed set of methods.[9]

A common method, however, is scoring specific, predefined indicators in interaction videos. For example, the success rate of a new algorithm for handover tasks between a robotic arm and a human participant can be measured by counting the number of successful events in a sample. Additionally, human–human interactions can be videotaped and evaluated in a quantified manner to model robotic decision trees and establish quantitative threshold values.[10] Common quantitative measurements extracted from video data may include the proxemic behavior of the participants in an interaction,[11] the duration and direction of gazing behavior,[12] de-

6   E.g., Christoph Bartneck/Jun Hu, Exploring the abuse of robots, in: *Interaction Studies. Social Behaviour and Communication in Biological and Artificial Systems* 9 (3/2008), 415–433.

7   Sarah N. Woods et al., Methodological Issues in HRI: A Comparison of Live and Video-Based Methods in Robot to Human Approach Direction Trials, in: *ROMAN 2006—The 15th IEEE International Symposium on Robot and Human Interactive Communication*, Hatfield 2006, 51–58.

8   Astrid Weiss et al., Autonomous vs. tele-operated: How people perceive human-robot collaboration with hrp-2, in: *Proceedings of the 4th ACM/IEEE International Conference on Human Robot Interaction—HRI '09*, 2009a, 257–258.

9   Louise Veling/Conor McGinn, Qualitative Research in HRI: A Review and Taxonomy, in: *International Journal of Social Robotics* 13 (2021), 1689–1709; Andreas Bischof, *Soziale Maschinen bauen: Epistemische Praktiken der Sozialrobotik*, Bielefeld 2017.

10   see Kyle W. Strabala et al., Towards Seamless Human-Robot Handovers, in: *Journal of Human-Robot Interaction* 2 (1/2013), 112–132.

11   See Jonathan Mumm/Bilge Mutlu, Human-robot proxemics: Physical and psychological distancing in human-robot interaction, in: *Proceedings of the 6th International Conference on Human-Robot Interaction—HRI'11*, 2011, 331–338.

12   E.g. Ajung et al., Meet me where i'm gazing: How shared attention gaze affects human-robot handover timing, in: *Proceedings of the 2014 ACM/IEEE International Conference on Human-Robot Interaction*, 2014, 334–341.

tailed space–time data about participants' hands in a handover situation,[13] or specific types of interactions or encounters with a robot, for example, abusive behavior,[14] and more. This type of analysis may include some interpretation, but it often remains—from a sociologist's perspective—on the surface of the data. The focus is not on the interpretation of interactions but rather on the classification of actions into a pre-given set of indicators to facilitate the quantification of those actions for mathematical modeling or statistical descriptions.

Since this way of doing research is considered to be the state of the art in HRI, it is obviously sufficient to inform a concrete design decision for building robots. However, such research is designed to answer very specific predefined questions and presumptions. This bears the risk of overlooking important aspects of situations and could preclude systematically reflecting upon preconceived contextual factors that influence the interactive techno-social phenomenon at hand. Therefore, such research is always liable to produce a reductionist view of complex processes. A more open and interpretive approach has its strengths in overcoming such problems and providing information about the entirety of socio-technical interaction. Of course, videos are neither the only nor the central methodological tool in the evaluation of experiments in HRI. We want to stress that, in this regard, our criticism has a limited scope and therefore cannot be generalized to all of HRI's methodological tool kit.

Although most HRI studies rely, as previously mentioned, on quantitative data, several studies have utilized qualitative methods. Veling and McGinn presented a great overview of qualitative research in HRI, and several of those studies used videos as an epistemic tool.[15] A couple of excellent studies have also used videographic methods, as we suggest.[16]

---

13   E.g. Matthew K. X. J. Pan/Elizabeth A.Croft/Günter Niemeyer, Exploration of geometry and forces occurring within human-to-robot handovers, in: *2018 IEEE Haptics Symposium (HAPTICS)*, San Francisco 2018, 327–333.

14   Dražen Brščić et al., Escaping from Children's Abuse of Social Robots, in: *Proceedings of the Tenth Annual ACM/IEEE International Conference on Human-Robot Interaction*, 2015, 59–66.

15   Veling/McGinn, Qualitative Research in HRI.

16   E.g., Morana Alač, Social robots: Things or agents?, in: *AI & Society* 31 (4/2016), 519–535; Karola Pitsch et al., Interactional Dynamics in User Groups: Answering a Robot's Question in Adult-Child Constellations, in: *Proceedings of the 5th International Conference on Human Agent Interaction*, 2017, 393–397; Karola Pitsch, Answering a robot's questions. Participation dynamics of adult-child-groups in encounters with a museum guide robot, in: *Réseaux* 220–221 (2/2020), 113–150; Antonia L. Krummheuer/Matthias Rehm/Kasper Rodil, Triadic Human-Robot Interaction. Distributed Agency and Memory in Robot Assisted Interactions, in: *Companion of the 2020 ACM/IEEE International Conference on Human-Robot Interaction*, 2020, 317–319; Antonia L. Krummheuer, Conversation Analysis, Video Recordings, and Human-Computer Interchanges, in: Ulrike Kissmann (ed.), *Video interaction analysis: Methods and methodology*, Frankfurt a. M., Berlin, Bern, Wien 2009, 59–83; Florian Muhle, Begegnungen mit Nadine. Probleme

In contrast to these examples and despite the ubiquitous use of videos generally, they are rarely used in HRI, and when they are, they are seldom employed within a distinct method(olog)ical approach—as the whole field lacks such approaches. The following excerpt explains this.

> HRI is a research area that remains young and highly interdisciplinary, and the approaches, standards, and methods are still in the process of negotiation. While this brings a high level of interdisciplinary attention, innovation, and creativity to the field, it also leads to challenges in establishing agreed upon systematic approaches and methods. The field has been widely criticized for its lack of scientific quality and methodological rigour.[17]

Presumably because of its ubiquity, video use is also rarely reflected upon. This has led to high variance in the video use typically found in HRI studies. Although statistical analysis and reflection are usually done very well and rigidly, there is room for improvement of the qualitative analysis of videos. We argue for the reflective and methodologically based use of video analysis to expand the HRI research toolbox.

We have found that doing so is especially crucial now that robots "leave their creators['] laboratories" (or rather are brought into human contexts) and enter "institutional settings" (see this issue). To understand why this is exceptional, it is worth recalling that the central site of knowledge production in robotics is the laboratory. As is the case for many other sciences, the laboratory is the place to "tame" worldly complexity and reproduce it in a controllable way.[18] For this purpose, complexity is first decomposed into processable units that are small enough to allow researchers to ignore complex dependencies.[19] It is the special environment within the laboratory that enables the functioning of the limited robotic technology at all. This is also true for HRI; although it is dedicated to the promise of everyday applicability, HRI exists primarily as a laboratory science.[20] This starts with basic technical aspects: In the lab, the environment can be controlled; light and sound quality can be kept stable (robotic labs' notorious blacking out of window is due to light control to allow for

der "Interaktion" mit einem humanoiden Roboter, in: Angelika Proferl/Michaela Pfadenhauer (eds.), *Wissensrelationen: Beiträge und Debatten zum 2. Sektionskongress der Wissenssoziologie* (1st edition), Weinheim, Basel 2018, 499–511; Florian Muhle, Humanoide Roboter als 'technische Adressen': Zur Rekonstruktion einer Mensch-Roboter-Begegnung im Museum, in: *Sozialer Sinn* 20 (1/2019), 85–128.

17   Veling/McGinn, Qualitative Research in HRI.

18   Bruno Latour, Give me a laboratory and I will move the world application, in: Karin Knorr Cetina/Michael Mulkay (eds.), *Science observed: Perspectives on the social study of science*, London 1983, 141–170.

19   Susan L. Star, Simplification in Scientific Work: An Example from Neuroscience Research, in: *Social Studies of Science* 13 (2/1983), 205–228, see 207.

20   Bischof, Soziale Maschinen bauen.

robotic vision, and videos profit from this), whereas in the real world, they fluctuate. Expedients such as markers on the floor to measure distances or guide participants help to implement standardized quantitative measures or even automatized metrics, but these are not available in institutional settings. Moreover, the opportunity to prepare participants is inapplicable because people cannot be preselected, instructed, or guided through as is common in lab experiments.[21] Often, interactions in the lab are completely staged using the so-called "Wizard-of-Oz"[22] method.[23]

None of this is true "in the wild," which demands that we be even more attentive to the methods to achieve the same significance as in the lab. Therefore, we propose transferring the methodological knowledge of videography[24] from social science and making it usable for HRI research in the lab, as well as—and especially—when studying robots "in the wild." In the following pages, we show that this type of qualitative analysis is a useful addition to the usual qualitative interpretation in HRI.

2. Opportunities for Videography in HRI Studies

Qualitative videography offers several opportunities for the analysis of HRI in institutional settings. This is, in part, due to the current state of the scientific field. HRI usually comprises complex interaction contexts.[25] Interactions with robots are also, due to the robots' technically immature nature, multimodally clumsy (which is why the scientific field of HRI is needed to deal with this problem in the first place). This, in turn, usually leads to multiple irritations for subjects that can—implicitly as well as explicitly—affect various aspects of the interaction (proxemics, language, etc.).

---

21   Benjamin Lipp, (2019). Interfacing RobotCare. On the Techno-Politics of Innovation, URL: http://nbn-resolving.de/urn/resolver.pl?urn:nbn:de:bvb:91-diss-20190624-1472757-1-8 [last accessed: August 15, 2023].

22   In an experimental setting, the "wizard", like his namesake in *The Wizard of Oz*, controls the robot invisibly to the test subject. The test subject expects the robot to act autonomously and not be controlled by an invisible human. The purpose of this is to compensate for technology that is not yet ready for use or to guarantee the robot's robust operation during experiments.

23   Astrid Weiss et al., User experience evaluation with a Wizard of Oz approach: Technical and methodological considerations, in: *9th IEEE-RAS International Conference on Humanoid Robots*, Paris 2009b, 303–308.

24   Knoblauch/Tuma, Videography; see also Heath/Hindmarsh/Luff, Video in Qualitative Research.

25   Andreas Bischof/Arne Maibaum, Robots and the complexity of everyday worlds, in: Benedikt P. Göcke/Astrid Rosenthal-von der Pütten (eds.), *Artificial Intelligence. Reflections in Philosophy, Theology, and the Social Sciences*, Paderborn 2020, 307–320.

For this reason, interactions with robots in particular are often characterized by ruptures, breached expectations, and other ambivalences and explanation-requiring behaviors. Those ruptures can be "repaired" (see Schegloff[26] for the concept of repair in speech; for its conceptual underpinning, see Garfinkel's 1967 work),[27] or alternatively, ignored by the human counterpart. For example, a robot to human handover might be successful because the human quickly adjusts their hand position to compensate for the robot's inaccuracy. Many HRI situations are novel for participants, so the learning effects are often very powerful and therefore play a strong role in longer-lasting interactions. Therefore, these interactions can often only be explained conclusively when observed in their entirety. This includes contextual factors of the situation that may not initially be in the analysis' field of attention but on which the analysis can be retrospectively focused. In this regard, very small details, such as actions related to repairing, or re-evaluating or negotiating the interactional order and similar human (re-)actions, can only be discovered in a thorough detailed analysis. Strong physical involvement also plays a more significant role, which makes it intriguing to include the situation's entire development, that is, from the first encounter onwards, in the analysis. Furthermore, in institutional settings, it is most probable that third parties will play a crucial role, which can subsequently influence interactions. Here, other—often overlooked—bystanders, such as an audience, the experimenters, or other participants, play a decisive role.[28]

Videography has a wider focus on social encounters and situations, and accounting for all these factors offers a perspective and a methodological tool kit that can address the more complex embeddedness in mundane and institutional settings. Processing data in this way is time-consuming but worthwhile.

3.1   On the Term "Interaction"

Before beginning our proposal on the use of videography to facilitate HRI video analysis, we will briefly probe the term "interaction" and its underlying concepts. A typical concern in the interdisciplinary field of HRI is that although HRI is often based in engineering departments, it systematically covers areas in the domain of the social and communication sciences, such as psychology and (multimodal) linguistics. This has resulted not only in the inclusion of humans but also in the adoption of key

---

26   Emanuel Schegloff, The Relevance of Repair to Syntax-for-Conversation, in: Talmy Givón (ed.), *Discourse and Syntax* (Vol. 12), Boston 1979, 261–286.

27   Harold Garfinkel, *Studies in ethnomethodology*, New Jersey 1967.

28   Diego Compagna/Claudia Muhl, Mensch-Roboter Interaktion—Status der technischen Entität, Kognitive (Des) Orientierung und Emergenzfunktion des Dritten, in: *Muster Und Verläufe Der Mensch-Technik-Interaktivität. Working paper: TUTS-WP-2-2012*, Berlin 2012.

concepts such as interaction, which takes on a specific meaning here. In HRI, "interaction" is commonly understood as a general term for the relations between two entities, although one of those might, in fact, be an object. In social sciences, the term "interaction" has a distinct meaning, referring traditionally to human subjects, even if approaches such as actor–network theory[29] challenge this. Without going into details and getting specific about the different uses of the term (which is also being used pragmatically in this paper), the process of interaction in social science studies is a complex process of reciprocal action that includes observation and anticipation, as well as the interaction partners' expectations and reciprocal conceptions[30] and future trajectories.

Depending on the conceptualization of interaction, in many traditions, the concept is reserved for bodily co-present situations and limited to human–human interaction. The paradigmatic case for social scientific video analysis is what Goffman has called "focused interaction," that is, the form of interaction in which bodily co-present participants share a common focus of attention.[31] In the simplest case, focus is constituted by two actors. However, focused interactions may also extend to larger social occasions, such as meetings, staged events, and demonstrations. This concept can be adapted for HRI interactions but might be extended by concepts such as the "synthetic situation"[32] for humanoid robots[33] or technologically extended situations.[34] Forms of interaction between technological agents and humans are then specifically referred to as "interactivity"[35] or described as "triadic interaction."[36] In the argument presented in this paper, we use the term "interaction" pragmatically

29   Bruno Latour, Reassembling the social: An introduction to Actor-Network-Theory (1st edition), New York 2005.

30   Alfred Schutz, Common-Sense and Scientific Interpretation of Human Action, in: Maurice Natanson (ed.), *Collected Papers I, The Problem of Social Reality*, Dordrecht 1962, 3–47.

31   Erving Goffman, Behavior in Public Places—Notes on the Social Organization of Gatherings, 1966.

32   Karin Knorr Cetina, The Synthetic Situation: Interactionism for a Global World, in: *Symbolic Interaction* 32 (1/2009), 61–87

33   Florian Muhle, Roboter in der Sozialwelt. Überlegungen und Einsichten zum Subjektstatus humanoider Roboter, in: M. Schetsche/A. Anton (eds.), *Intersoziologie: Menschliche und nichtmenschliche Akteure in der Sozialwelt*, Weinheim, Basel 2021, 128–142.

34   Eva Hornecker et al., The Interactive Enactment of Care Technologies and its Implications for Human-Robot-Interaction in Care, in: *Proceedings of the 11th Nordic Conference on Human-Computer Interaction: Shaping Experiences, Shaping Society*, 2020, 1–11.

35   Werner Rammert, Where the Action is: Distributed Agency between Humans, Machines, and Programs, in: Uwe Seifert/Jin H. Kim/Anthony Moore (eds.), *Paradoxes of Interactivity* (1st edition), Bielefeld 2015, 62–91

36   Antonia L. Krummheuer/Matthias Rehm/Kasper Rodil, Triadic Human-Robot Interaction. Distributed Agency and Memory in Robot Assisted Interactions, in: *Companion of the 2020 ACM/IEEE International Conference on Human-Robot Interaction*, 2020, 317–319

and actively include encounters between humans and robots as long as they are embodied and experienced as robots.

3. Suggestion for a Methodological Reflected Use of Qualitative Video Analysis in HRI

Based on this understanding of interaction, we describe three typical examples for the specific use of qualitative video data in HRI. We understand these not as instructions but rather as entry points to reflect on the specifics of videographic methods for HRI. Before diving into the examples, we will briefly make some general remarks on videography as a methodological component and identify the most important aspects regarding HRI contexts.

In the process of videographic research, the first central question is usually what to focus the camera on and how. This initially seems like a simple question, but there are numerous factors to be considered when answering it. The question of focus is, of course, dependent on the research questions, which should be informed by theory, the research area, and the discipline. Generally, if focusing on an interaction, the relevance can, on the one hand, be set by the observer's specific question, or on the other hand, follow the relevance of the participants "in front of the camera." To give an example, although instructions or audiences in a given experiment situation may not, at first glance, be relevant to the analysis of a specific robotic movement, they might be crucial to the human participants' actions, and it is helpful to have those defining moments recorded for later interpretation. This is not only to address "mistakes" but also to open up the possibility of understanding participants' resources to make sense of the situation. This becomes quite clear not only when selecting events to be recorded but also in relation to the modalities on which we focus: In the tradition of video analysis in the social sciences, which stems from conversation analysis, the strong focus has always been on the spoken language.[37] However, in HRI, this relation is reversed: In most HRI studies, the relevant interaction is a visual–tactile interface. We suggest that this needs to be recorded, or if an interaction is based on moving bodies in space, this dimension should be the general focus, without neglecting the other. Even if one is only interested in a specific movement, for example,

---

37   In recent years, however, linguistics has closely examined aspects of what are termed "modalities": Next to lexical choices, codes and prosody, gestures, facial expressions, or body posture, an interdisciplinary field has emerged, populated by researchers who are strongly influenced either by structural linguistics or by the more recent pragmatic cross-fertilizations between linguistics and neighboring disciplines (For an overview see Tuma et. al 2013 chapter 3, see also Adam Kendon, The F-formation system: The spatial organization of social encounters, in: *Man-Environment Systems* 6 (1/1976), 291–296; Mondada, Multiple Temporalities of Language and Body in Interaction).

the accompanying talk might be crucial, and vice versa. HRI is often less interested exclusively in spoken interaction, focusing instead on visual and tactile interaction.

In complex settings, where interactions occur across a range of modalities, all of these should be recorded, as they add to the complexity of sensemaking—not only for participants but also for researchers. This also profoundly influences which recording technique should be used: If, for example, the robot is moving in a large common room at an elderly care home, both the situation as a whole, with all actors involved, as well as the spatially close interactions between the elderly and the robot, must be recorded. A simple solution for unobtrusive recording could be using the robot's visual sensors or placing an additional camera in or on the robot. Under some circumstances, it may be necessary for the camera to pan or tilt to follow the participating observer's gaze. In controlled experimental or laboratory settings, focusing on the experiment is a clear task, as the space is built for the occasion; however, identifying relevant actions and the factors that might influence the observed encounter is not always straightforward for the researcher "in the wild."[38]

It should be made clear that HRI situations are usually constructed as very specific social events that result in a specific complexity. Those complexities need to be addressed, but they also offer resources for generating further insights. Therefore, we will, using examples, highlight the most relevant distinctions to adequately describe this complexity. Since researchers both actively reduce and actively increase an HRI situation's complexity,[39] it is more instructive to speak about dimensions that increase or decrease the degree of control over the scripted interaction. In the following pages, we present typical examples of video-based HRI studies based on the experimenters' level of control. Of course, the list is not exhaustive, as there are other dimensions that matter, for example, the participants' degree of familiarity with the technology, their awareness of the experimental situation, and more. Since we cannot touch upon all of them, we focus on the level of control, which can be regarded as the most important when addressing the particularities of institutional settings. We have sorted the examples in descending order of the degree of control exerted over the situation, starting with the laboratory setting.

---

38   For HRI, this is more complex, as there are some specifics to this form of interaction. Historically, video interaction analysis was established for the analysis of human–human interaction. Therefore, interpretation aims to understand what human participants have done, and we can (to a certain degree) use our own experiences as mundane members of society to understand what typical observable bodily and spoken turns or actions in a situation "mean." However, we refrain from far-reaching "motivational" interpretation and stick with "situational" interpretation based on the embedding in a meaningful sequence. For example, a specific "grabbing" gesture can be understood as an "intention of reaching out for an object," but we refrain from interpreting it as an expression of motivations such as greed, general neediness, or other possibilities.

39   Bischof/Maibaum, Robots and the complexity of everyday worlds.

4.1   Setting 1: The Fully Controlled Lab Environment

We start with examples that show typical laboratory settings as the most controlled experimental HRI setup. These closed laboratory settings allow the roboticist to control all factors influencing the situation in general and the robotic system in particular. The fully controlled laboratory experiment is the most widely used setting in HRI research to date. The focus is on clearly defined tasks of a robot in interaction with a human counterpart. The complexity of the experiments can vary widely, but typically, all relevant interactions occur in a certain limited time frame guided by a task at a specific location. The material environment at this location is controlled and adapted to the experiment's needs: Tracking systems are installed, furniture is placed, items are prepared for interaction, and markers are positioned on the floor. The closed laboratory also allows for control of the sample of participants. Participants are selected based on desired characteristics; for example, participants who are unfamiliar with the robot may be recruited. Before the experiment, participants may be instructed and informed about the robot's alleged or factual capabilities or shortcomings.

These laboratory experiments are often videotaped as such closed experimental spaces allow recordings conveniently. The local infrastructure and the stationary experiments allow fixed camera positions; for example, cameras may be mounted inconspicuously on the ceiling or hidden in the chaos of the lab. Such camera setups often also have the side effect of reducing participants' reactance to being recorded. As can be seen in Figure 1, these videos are often used to demonstrate seamless interactions or well-functioning robotic technology and therefore focus on the direct concrete point of interaction and/or the robotic system.

Consequently, these forms of experimental recording pose some challenges for use in qualitative analysis: It can easily happen that the recording's focus is too narrow on the interaction itself, overlooking other contexts that are also important for the functioning of the situation. The control exerted in a laboratory setting can therefore have a deceptive effect, requiring special attention to the circumstances, especially in the case of more complex questions. In the analysis, it is important to address what impressions and information participants were exposed to through the process and the framing of the robot when interacting with it and what influence these might have had; that is, the question of how the participants interpreted the situation (e.g., as a game, as playing with technology, as a serious demonstration, etc.) must be answered.

Fig. 1: A camera out of sight of the participant and a direct top view from the ceiling. The video is used to demonstrate the capabilities of the robot rather than get information about the interactions

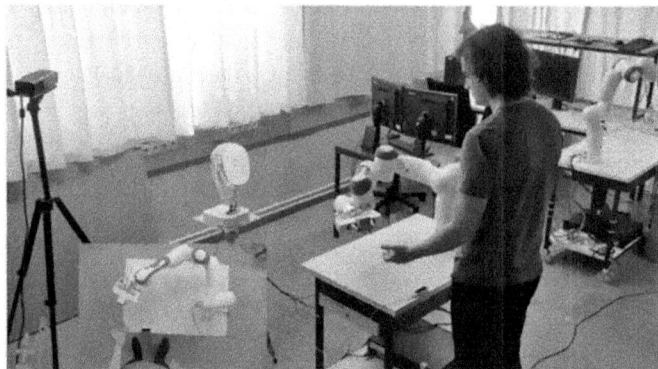

Therefore, in a laboratory experiment, it is always important to keep in mind that participants are likely aware that they are being observed.[40] Recording the reasonably widest context of the setting, which covers the whole room, all actors involved (including the HRI researchers), and the complete process in a total perspective is an attempt to facilitate an analysis that can provide insights into the participants' entire range of experiences of the situation. Decisions on the camera position, the angle chosen, and the modalities of the record but also the timing of the record have to be made carefully because they determine the possibilities of the analysis of the gathered data, as we will show hands-on with example pictures depicting a typical HRI situation: a handover. A typical robot task in experiments are handovers, during which the robot is supposed to hand over an object (often a soft ball, as in our example) to a receiving human. This mimics future uses such as handing over tools or manufacturing parts.

Figure 2 shows an extreme but not completely uncommon perspective on the handover task, where only the directly relevant body parts are recorded, not the whole person or the whole robot, which excludes the human's gestures, facial expressions, and body postures. Moreover, the timing of the record here is pragmatically bound to the small sequences of direct interaction, thereby missing important side aspects happening before and after.

---

40   Susan A. Speer/Ian Hutchby, From Ethics to Analytics: Aspects of Participants' Orientations to the Presence and Relevance of Recording Devices, in: *Sociology* 37 (2/2003), 315–337.

Fig. 2: A narrow perspective that only shows the point of interaction.

Fig. 3: This perspective also shows the upper body, but not the foot position and the remaining structure of the robot.

Fig. 4: This perspective shows the full situation as is centered on the point of interaction strictly from the side.

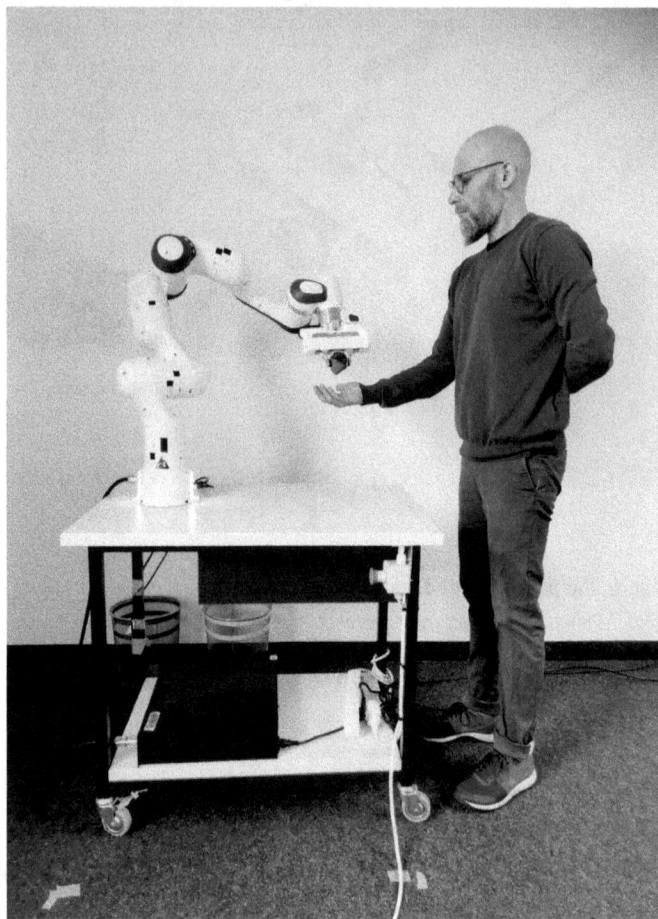

Choosing a broader angle of view of the same situation (see Fig. 3) opens up the possibility of capturing the mimic of relevant parts of the person and the whole robot; nevertheless, information is still missing here, for example, the feet as part of the person's whole body stance—information that is relevant to reconstructing the proxemic conduct. Additionally, possible markers on the floor or other actors present in the situation that might have exerted an influence are not getting recorded here. Figure 4 includes those aspects by capturing the "whole bodies" of the interacting human and the robot, but because it captures the situation strictly from the side, actors' distance information is hard to grasp, and the participant's face is not fully visible. A high view over the robot's "shoulder", as shown in Figure

5, captures most of the participant's actions in the situation and is therefore often the optimal camera perspective.

Fig. 5: A perspective from the top and from the side provides the best view of the participants.

Fig. 6: A wide angle view of the example situation covers the 'wizard' of the situation as well as the contextual objects and the placements of the interaction space.

However, it is not only the immediate proximity of the experiment space that significantly influences the experiment's outcome; an HRI experiment usually spans the entire room. A crucial role is played, for example, by the experimenter or the wizard, who remains—often intentionally—outside the direct interaction, so, for example, instructions, beforehand or in situ, ruptures in the interaction that the experimenters repair, or their provision of active assistance during experiments to correct or repair errors as they occur might only be seen if the interaction is recorded. An additional wide-angle perspective covering the whole room, as in Figure 6, can add information about those processes and reveal otherwise unseen factors. It might be necessary to document the technical infrastructure that may also be hidden, possibly even in another room. Especially in cases where a wizard takes control of the robot, the operating human is part of a translocal and mediated situation.

To get an overview of all potentially influencing factors, the wide-angle camera should record the experiment's complete time frame including preparation, instruction, and first contact with the robot. Additionally, the participants' actions before and after the experiment may contain relevant information. In experimental setups, the influence of experimenters or other side factors is often considered erroneous, and work is put into minimizing such experimental effects/artifacts; however, a detailed analysis of *how* contextual aspects shape specific interactions gives important insights that are often overlooked.

4.2   Setting 2: Not Fully Controlled Settings

The second setting is "semi-wild," where an experimental setting is embedded in a public space. Figure 7 shows a public demonstration of the pressure-controlled robotic arm BROMMI:TAK as part of the Long Night of Science event held in Berlin in 2015. The centrally located courtyards attracted a large audience that gathered around the robot.

Fig. 7: Overview of the not fully controlled example during a science event at the TU Berlin. The image shows the audience (1), the robotic arm (2), the wizard (3) and the experimenter (4).

Fig. 8: A typical video recording for interaction from a perspective close to the handover. Here you can hardly see the audience.

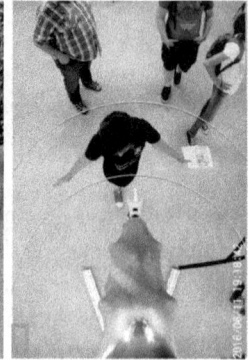

The spectators were interested in learning about the roboticists' innovative technology, but the team of HRI researchers were trying to find out how people would make sense of the robot's appearance and behavior. The participants were told that the robot would soon choose one person to give an apple, and the chosen person's task would simply be to receive the apple. Because the robot was controlled by a wizard, there was no specific robotics research interest in the setup; the team assumed that the robot's spatiotemporal behavior would sufficiently inform the participants of the identity of the chosen interaction partner.[41] Two roboticists on the team engaged in constant conversation with the event's coming and going attendees, explaining the robotic system and organizing the demonstration.

Despite this experiment's public nature, the recordings often focused on one close participant and the robot, with only the nearest bystanders included in ad-

---

41   As we will see in a later example, this was not the case, and the audience played a crucial role in selecting the interaction partners.

dition to those. For example, Figure 8 is a classic top-down perspective on a public event featuring the same robot, where the robot and a single handover participant are pictured. The biggest difference compared to the first setting is that the fully controlled laboratory space has been left behind, and the experiment has had to be integrated into a *different social* context. As can be seen in Figure 7, researchers in such settings only have limited control over the types and number of people interacting with the robot, as they are at an event that is open to the public. This affected the sampling of potential interaction partners for the observation, obscured the participants' prior knowledge, and also introduced material constraints given the technical system's higher error-proneness. It can be assumed that some of the participants suspected that they were part of an observation, but that was not necessarily the case, nor did it have to become relevant. However, as demonstrated in the overview, the situation is characterized by the fact that a wide variety of people or groups of people whose prior experience with robots was difficult to assess interacted with the robotic arm at irregular intervals. They were participants in an experiment but simultaneously acted in other social roles, for example, as parents, friends, or partners. Additionally, the influence of the roboticist leading the situation cannot be underestimated, as he instructed the interaction partners regarding the robotics system, with which they were likely unfamiliar. At the same time—and in opposition to Setting 1—the roboticist could not always sustain the same framing of the situation for two reasons: First, he was under pressure to fulfill the expectations of visitors who had come to see the demonstration and/or successfully interact with the robot; and second, the situation had a fluctuating audience and no sequential order, making it difficult for the experimenter to use precise wording when explaining the experimental setup.

Because such settings involve a large group of people who do not know each other and are participating in a probably clumsy interaction with an unfamiliar robotic prototype, felt uncertainties increase. From the participants' perspective, the situation is more complex to assess, as there are more actors present in the situation that may become relevant for their interaction with the robot.

These aspects present severe challenges for recording.[42] Consequently, it may be necessary to point two cameras at the direct points of interaction to capture as many details as possible for a reliable videographic analysis (see Fig. 8 for an overview of the situation). Videos that include both camera angles side by side, as in Setting 1, can be used to access the often multilayered interactions. Figure 4 shows an example of such a case. The robot is again supposed to pass an apple to one audience member, as the wizard moves the robotic arm. The wizard reacts to participants' conduct, which he can observe through a live recording, which means that the interaction is

---

42   Including acquiring formal (GDPR) consent from audience members who are influencing the situation but are not active experiment participants.

initiated by the participants themselves rather than by the wizard. Child 1, standing on the right side, just received an apple, and the audience's attention now turns to Child 2, standing on the left side. Child 2 is eagerly awaiting selection but does not step forward or reach out his hands to cue the wizard. The two angled shots shown in Figure 9 reveal the situation's complexity. Not only does Child 2 get pushed into the interaction by the girl behind him, but the older person on the right—visible in the second angle—takes a role in the situation as he instructs the wizard in the background by pointing at the child. Moreover, the man on the left side influences the situation when he uses his gaze to encourage the older man to ensure that the child wearing the cap is selected.

Fig. 9: The second camera angle (2) gives a clear view of how Child 2, standing on the left, gets pushed to interact with the robot.

The actions of individuals in large groups of people call—even more so than in Setting 1—for an overview perspective, as shown in Figure 7. In so doing, conclusions can often already be drawn, and interpretations of a person's actions are informed before the interaction with the robot starts. For example, the way a person approaches the experiment situation and how and when they decide to participate may significantly impact their interaction with the robotic system and its capabilities, which, in turn, manifests in the interpretation of the data, where these effects can be identified.

4.3   Setting 3: "Real-World" Institutional Work Settings

The third setting is an actual "in the wild" (or, as roboticists put it, "in the field") HRI "experiment," where a robot is tested and evaluated under quasi real-world conditions. The main characteristic—compared to Settings 1 and 2—is the avoidance of an experimental character that should optimally fade in the background. The purpose of such a field setting can be to demonstrate the robot's robustness or evaluate its interaction with uninstructed or unfamiliar persons. Two forms of "in the wild" contexts can be distinguished. On the one hand, there are public situations, for example, at a train station or on the street, in which a robot is supposed to interact with changing and always new interaction partners who pass by the scene. Here, the situational context should be generally accessible, which also makes it possible to rely on common knowledge to analyze the data. On the other hand, there are situations in which robots are integrated into institutional contexts, for example, a care home, where they are supposed to function alongside existing work practices and routines. The most important difference between these two is the extent to which visitors recognize the experimental character and the time frame of the interaction, as well as the extent to which the researchers need special knowledge of the field of expertise to make sense of the actions observed. Normally, in public settings, the experimental character remains hidden (to a certain degree), and the interactions are one-off and short term. In institutional work contexts, however, participants interact repeatedly over a longer period and adapt to the "robotic intervention" in their daily business,[43] causing the experimental character to fade into the background. However, it can never be ruled out that interaction partners are aware that they are part of an experiment.

From the viewpoints of both the experimenters and the participants interacting with a robot, this is the most complex setting. Here, the most uncontrollable material and social factors are to be expected: Even if the material environment is known

---

43   This may also lead to neglect of the robot, avoidance practices, or very selective "staging" on the part of the researchers in some cases, which can, however, be a resource to uncover the reasons for it.

to the robot, the system must meet the human interaction partners' social and situationally changing expectations. They, in turn, will most likely show no mercy to its failures because they are relying on the robot's actions to get their work done.

Such a setting is depicted in Figures 10–12, where a cat-shaped vacuum cleaner robot is being tested in an elderly care home in Denmark.[44] The study was designed to investigate the extent to which an abstract cat-shaped shell can increase vacuum cleaner robots' acceptance in care homes. The research also inquired as to whether playful movement can help decrease elders' anxiety about interacting with robots and increase their motivation to do so. The robot's movement in this setting was realized by a mobile wizard, who was also present in the room, remotely controlling the robot's movements with a gaming controller. In this example, we find a mixture of both forms: Although the staff knew about the experiment, the nursing home residents, some of whom were suffering from severe forms of dementia, did not always recognize that they were part of an evaluation.

Fig. 10: Total view of the elderly care home common room.

---

44   The example is taken from a study associated with the ReThiCare project. The subproject was realized by Emanuela Marchetti.

Fig. 11: Hand camera view of the room from the   *Fig. 12: Robot view of a situation, in which the*
other side. On the left side of the picture, the mobile *robot approaches an elderly person.*
wizard can be seen.

Such a setting comes the closest to videography's intention to observe humans' natural[45] or real-world actions. If a recording takes place "in the wild," the focus of the interaction is very often not only an HRI but also a social event, where a robot is participating. Therefore, before singling the HRI out for study, it is important to understand what the human participants do in such a situation, what their relevancies are, what kinds of plans/trajectories they follow, etc. In the example of the elderly care home, the researchers chose different recording perspectives to flexibly capture most of the actions in the situation. A room total view was used to get an overview of at least half of the big room. Most of the recordings were captured by a researcher who dressed as a caregiver (to attract less attention from the elderly residents) and followed the robot using a handheld camera. Additionally, a camera mounted on the robot recorded the interactions happening directly in front of the robot while it moved around the common room.

Using a handheld camera has certain advantages, but it also has pitfalls. For example, the researchers can flexibly follow the interaction and freely change the perspective if necessary, for example, if a new person enters the room. Of course, this method always involves the risk of choosing the wrong focus in a particular situation. The use of a roving camera has advantages, but this technique might also shape the situation because participants will react to this visible act of observation. HRI often takes place in situations that require participants' full attention; therefore, the presence of a handheld camera can fade into the background, enabling a close and "natural" view of the unfolding actions. When a robot is present, it is likely to attract sufficient attention to itself to support this process. Utilizing the robot's perspective or, as in the present example, mounting an additional camera on it may be an

---

45   At the same time, however, it must be emphasized that due to the novelty of robots in society, there are yet no truly "natural" interactions with robots, as these are usually initial encounters that do not include long-term interactions.

additional recording feature that enables researchers to come very close to the interaction without disturbing it. The "wizard" in such settings (if used) plays a crucial role, as they are in charge of staging the correct behavior on the part of the robot. Therefore, it may be advantageous to also record these actions—but that is a hard task if the wizard is moving around. In such a case, it may be mandatory to include the wizard in the interpretation process, which may shed light on the wizard's situated interpretations during the interaction.

When using an autonomous robot, process-generated data may constitute an even more important data source to inform interpretation. A good example of the complexity and outcomes of studies conducted "in the wild" are those on autonomous driving, where it is very helpful not only to have a static understanding of traffic but to study participants' mutual interpretation. Many vehicle sensors can provide data for subsequent (meaningful) analysis (Fig. 13), but these might be insufficient to understand how human participants interpret the vehicles' "actions" (see Albrecht [in preparation])[46] and how those interpretations form part of the interaction's trajectory.

In such settings, however, the idea of achieving "total recording" is unrealistic, which is why we emphasize the importance of systematically combining fine analysis with other knowledge sources, especially ethnography.

Fig. 13: Example for a study on autonomous driving (See Albrecht in preparation)

---

46   F. Albrecht (in preparation), Technology assessment on autonomous vehicles: The routine grounds of traffic interaction, in: Singh, Ajit; zu Verl, Christian, Tuma, René: Videoanalysis in action.

4.4   Ethnography's Important Role for Videography

Given our examples' increasing complexity, it is pertinent to emphasize that the amount of interpretation necessary to make sense of the data has been increasing as well. Consequently, we suggest a methodologically reflected and integrated use of the recordings. For this reason, we recommend refraining from restricting the research to the analysis of videos and instead using videographic analysis, that is, video-based, *and* ethnographic analysis as a method.

Ethnographic methods typically include observation, some form of participation in the field, interviewing, eliciting, and collecting field documents. These methods allow researchers to gather subjective knowledge of the field under investigation, that is, to do the ethnography before the video interaction analysis, which allows for an understanding of other participants in the field. As Heath and Hindmarsh have pointed out, in the analysis of video recordings of naturally occurring activities, "it is critical that the researcher undertakes more conventional fieldwork."[47] For routine research practices, this implies that it is most useful if the researchers doing the video analysis are the same ones that began with the ethnography.

In videography, contextual knowledge is systematically collected and combined with the video-based analysis of situative "turns of action" as scrutinized in detailed video interaction analysis.[48] Ethnography is helpful prior to recording, as specific demands on how to produce the video recordings are revealed only by the ethnographic work that precedes it. Therefore, the ethnographic part allows for the context and the "backstage" of the recorded settings to be addressed. The ethnographic preliminary work not only helps to identify the specific situations to be selected for video analysis, but also to uncover the underlying situational and institutional contexts as well as the subjective knowledge of the participants and researchers.

We previously touched on the question of whether interaction with robots is interaction in the sociological sense. From a conservative sociological perspective, there are good arguments against such a view, for example, the fact that robots are not comparatively knowledgeable actors who have expectations and expect expectations from their counterparts. In the sociological sense, a robot (as things stand) can in no way be understood as an alter ego to which similar abilities and intentions as oneself could be attributed—at least, not if disappointed expectations are to be avoided. HRI is well aware of this problem, and therefore, most of the time, researchers attempt to reduce contingencies in experimental interactions

---

47   Christian Heath/Jon Hindmarsh, Analysing Interaction: Video, Ethnography and Situated Conduct, in: *Qualitative Research in Action* (2002), 99–121.

48   See René Tuma/Bernt Schnettler/Hubert Knoblauch, Videographie. Einführung in die interpretative Videoanalyse sozialer Situationen, Wiesbaden 2013., see chapter 6.

with robots.[49] For the research practice of video-based interaction analysis, however, this circumstance creates a double problem. On the one hand, the researchers themselves may lack knowledge of the robot's capabilities and functions, making "interpretation" of its actions essentially impossible. Regarding interpretation with respect to robotic participants, we do not have the "natural" resource of our own experiences as participants available to interpret the robot's "behavior" since their workings differ from human actions. An interpretation of a robot's behavior can only entail reading sensor values and following decision trees, and doing so has little in common with the interpretation of human conduct and intention. However, these are often available along with other resources, such as specific codes, protocols, or—most importantly—the roboticist experimenter's knowledge, which can all complement the material for interpretation.

On the other hand, the problem of interpretation is doubled by the fact that the interaction partner in an experiment also lacks this natural resource. This makes interpreting this actor's interaction with a robot challenging. Although we can draw on our own experiences of using various strategies to deal with deviant actors, there are many possible attributions, interpretations, or mental models that a person can attribute to the robot's behavior. To make matters even more complicated, the attributions made are often tested, revised, and rebuilt as the situation progresses. This also shows how important it is to not only look at the situative interaction but also to keep the technical infrastructure of an HRI setting in mind because it shapes and scripts the situation (see the laboratop).[50] In many cases, the robots are not autonomous; they are monitored and controlled by other humans (see above examples), who are an essential part of the situation that should be recorded. In addition to the observer's general competence in understanding others' actions, video analysis presupposes another kind of knowledge that is generally termed "ethnographic."[51] Since the structure of knowledge in modern society is highly specialized

49   Andreas Bischof, The Challenge of Being Self-Aware When Building Robots for Everyday Worlds, in: Athanasios Karafillidis/Robert Weidner (eds.), *Developing Support Technologies*, Heidelberg 2018, 127–135.

50   Juliane Haus, *Das ökonomische Laboratop: Eine soziologische Ethnographie des wirtschaftswissenschaftlichen Experimentierens*, Wiesbaden 2021.

51   One should note that definitions of "ethnography" differ significantly across disciplines. The type of ethnography to which we refer, namely focused ethnography, typically does not aim at encompassing large, locally distributed social structures, such as tribes or villages. As opposed to such encompassing "conventional" ethnographies (as we call them for the sake of brevity), for HRI, we focus on focused ethnography, which utilizes short but well-prepared visits and produces dense data through fieldwork (Hubert Knoblauch, Fokussierte Ethnographie: Soziologie, Ethnologie und die neue Welle der Ethnographie, in: Sozialer Sinn 2 (1/2001), 123–141.

and fragmented into many different settings, situations, and institutions, social scientists are often *not* familiar with aspects of social situations, for example, the setting in a (specific) robotic laboratory.

Nowadays, recording time is usually not limited by technical means (storage space) but rather by specific selections in the field. It is possible to achieve full-time, total recording in lab settings, but when recording "in the wild," it is rather tedious to capture all the important influences that define the situation. Therefore, the various aspects are often covered not by trying to capture a full recording of the complex social occasion but rather through ethnography and the combination of data sources. As our examples have shown, HRI is not interested in understanding "how a lab works" or in all the aspects of the "public display of technology"; rather, the questions are usually more focused, which leads to the selection not only of situations to record but also to the sampling of sequences out of the recorded data. It has proven to be helpful in ethnographic research to create some form of logbook or index of what is recorded in the data, including notes on interesting events and the course of action taken—basically, a protocol of what happens in the data that should be done rather openly. Including aspects such as "side talks," "failures," "phases of setting up," and "external irritations," with some details, might become important later.

4.5   Selecting Sequences and Time Units

Based on the research question at hand, the data are then screened for relevant sequences, such as successful instances of handing over an object. For a sample of such sequences, a detailed transcription is produced (Fig. 14), which is an essential step in the analysis process, although it allows for the interpretation of meaningful action rather than just "technically" tracking movements. For videography using the tools of conversation analysis and multimodal analysis, a very fine-grained transcript including spoken language (including pauses, pronunciation, etc.) is produced, then enhanced by aptly transcribing bodily movement as well. For HRI, because bodily movement is not as dominant in the interaction, other means of transcription (in addition to directly working with the video) are used, such as multimodal transcription,[52] which enhances classical transcripts,[53] or other ways that adapt to the specific situation and often rely on visual forms.[54]

Fig. 14: Example cut of the transcript with attached redrawn screenshots from the video-recording.

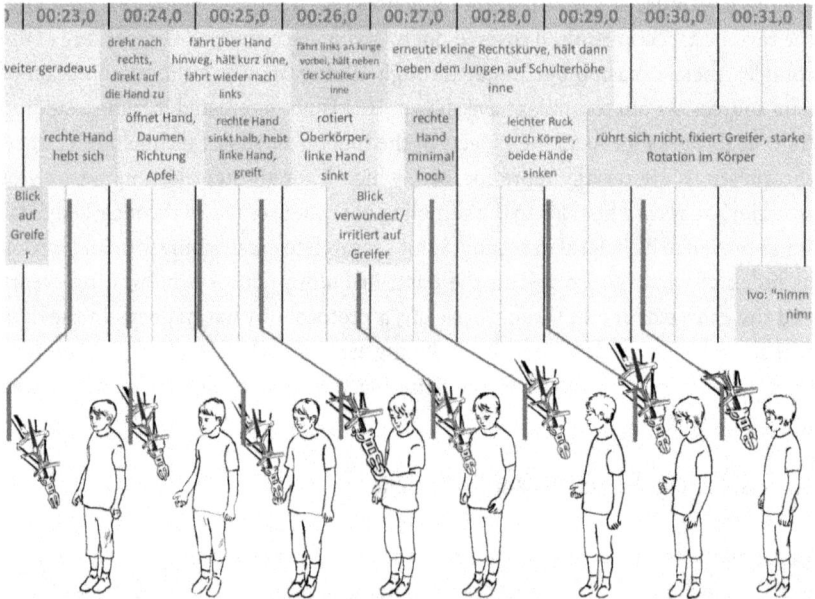

The sequence's duration is guided by two aspects: first, the research question, that is, what is relevant to answering the core theoretical question, and second, connected to that, the specific actions the participants perform. Meaningful units are not only observed but produced and marked by the actors in the video, for example,

52   Lorenza Mondada, Multiple Temporalities of Language and Body in Interaction: Challenges for Transcribing Multimodality, in: *Research on Language and Social Interaction* 51 (1/2018), 85–106.

53   For classical transcription systems, see Gail Jefferson, Glossary of transcript symbols with an introduction, in: Gene H. Lerner (ed.), *Conversation Analysis: Studies from the First Generation*, Amsterdam 2004, 13–31, for video transcription: Tuma/Schnettler/Knoblauch, Videographie; Christian Heath/Jon Hindmarsh/Paul Luff, Video in Qualitative Research: Analysing Social Interaction in Everyday Life, London 2010.

54   Saul Albert et al., Drawing as transcription: How do graphical techniques inform interaction analysis?, in: *Social Interaction. Video-Based Studies of Human Sociality* 2 (1/2019); Eric Laurier, The panel show: Further experiments with graphic transcripts and vignettes, in: *Social Interaction. Video-Based Studies of Human Sociality* 2 (1/2019).

whether a successful object handover has been achieved; then, there will be confirmation, followed by initiation of a new action. The researcher can use markers for those beginnings and endings of an interaction sequence to limit the scope of the analysis and find contrasting cases. This might sound simple in rather standardized situations, but it can be more complicated in complexly intertwined and embedded interactions "in the wild," where participants do several things at once or follow longer trajectories. The identification of sequences might sound quite basic, but it provides a starting point for finding patterns in interactions and comparing them to similar as well as to deviant cases. For example, if several instances of "handing over an object" can be identified, including how participants initiate and end this "task," it might be helpful to look especially for instances where not only does the task fail but other aspects, such as the initiation, also fail, or the sequence is interrupted. Problematic cases are a particularly rich resource for identifying the important aspects of the "normal" routine of doing things.[55]

Fig. 15: The videographic research process for HRI[56]

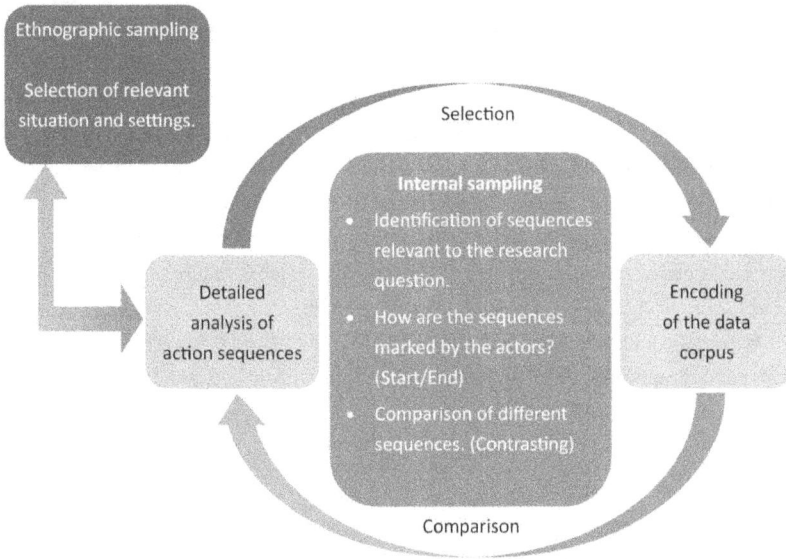

The systematic use of a larger corpus of data, the selection of cases based on theoretical interest, and the construction of typologies and systematic comparisons

---

55   C.f. Garfinkel, Studies in ethnomethodology.
56   See Tuma/Schnettler/Knoblauch, Videographie.

allow for the generation of insights beyond a single instance or case. With such a perspective, empirical research is always based on a theoretical endeavor (e.g., by analytic induction).[57] For social scientific research on robots, such a perspective allows for the embedding of practical research into theoretical frameworks, for example, the sociology of knowledge,[58] interaction concepts,[59] science and technology studies,[60] or new materialism.[61] Figure 15 visually synthesizes the suggested videographic research process for HRI.

4. Conclusion

Our paper aims to address and improve typical uses of videos in HRI research and demonstrate the potential of using social science-informed videographic methods. Although the use of video recordings is common and widespread in the field of HRI, there is little consensus on methodological approaches to video analysis and often no reflection on them at all. We saw this situation as an opportunity to suggest the use of distinct videographic methods for analyzing videos of HRI, especially in situations outside the laboratory, namely institutional settings, that differ substantially from laboratory settings. We have compared the settings, especially regarding the level of control that experimenters have over the context factors of the experiment situation.

To illustrate the method's advantages and give insights on its use, we drew on examples from our research. We followed the complexity of the situation analyzed from an isolated experiment in a laboratory, through semi-controlled settings, all the way to research "in the wild." Following these suggestions, we elaborated on the importance of ethnography for videographic work in HRI and asserted that ethnographic knowledge is not only necessary to make sense of the recorded data but is also crucial for the conception of the video recordings.

We have highlighted that this means a shift of perspective from isolated interactions towards a more holistic or relational understanding of embedded interactions

---

57 Jack Katz, Analytic Induction, in: Neil J. Smelser/Paul B. Baltes (eds.), *International Ency-clopedia of the Social & Behavioral Sciences*, Oxford 2001, 480–484.

58 Michaela Pfadenhauer/Christoph Dukat, Robot Caregiver or Robot-Supported Caregiving?. The Performative Deployment of the Social Robot PARO in Dementia Care, in: *International Journal of Social Robotics* 7 (3/2015), 393–406.

59 Roger Häussling, Video analysis with a four-level interaction concept: A network-based concept of human-robot interaction, in: Ulrike Kissmann (ed.), *Video interaction analysis: Methods and methodology*, Frankfurt a. M. 2009, 107–31.

60 Lucy Suchman, *Human–Machine Reconfigurations: Plans and Situated Actions* (2nd edition), Cambridge 2007.

61 Lipp, Interfacing RobotCare.

that are becoming increasingly important as HRI enters more and more mundane and institutional "in the wild" settings.

Bibliography

Alač, Morana, Social robots: Things or agents?, in: *AI & Society* 31 (4/2016), 519–535.

Albert, Saul/Heath, Claude/Skach, Sophie/Harris, Matthew T./Miller, Madeline/ Healey, Patrick G. T., Drawing as transcription: How do graphical techniques inform interaction analysis?, in: *Social Interaction. Video-Based Studies of Human Sociality* 2 (1/2019).

Albrecht, Felix (in preparation), *Technology assessment on autonomous vehicles: The routine* grounds of traffic interaction, in: Singh, Ajit; zu Verl, Christian, Tuma, René: Videoanalysis in action.

Bartneck, Christoph/Hu, Jun, Exploring the abuse of robots, in: *Interaction Studies. Social Behaviour and Communication in Biological and Artificial Systems* 9 (3/2008), 415–433.

Bischof, Andreas, *Soziale Maschinen bauen: Epistemische Praktiken der Sozialrobotik*, Bielefeld 2017.

Bischof, Andreas, The Challenge of Being Self-Aware When Building Robots for Everyday Worlds, in: Athanasios Karafillidis/Robert Weidner (eds.), *Developing Support Technologies*, Heidelberg 2018, 127–135.

Bischof, Andreas/Maibaum, Arne, Robots and the complexity of everyday worlds, in: Benedikt P. Göcke/Astrid Rosenthal-von der Pütten (eds.), *Artificial Intelligence. Reflections in Philosophy, Theology, and the Social Sciences*, Paderborn 2020, 307–320.

Brščić, Dražen/Kidokoro, Hiroyuki/Suehiro, Yoshitaka/Kanda, Takayuki, Escaping from Children's Abuse of Social Robots, in: *Proceedings of the Tenth Annual ACM/ IEEE International Conference on Human-Robot Interaction*, 2015, 59–66.

Compagna, Diego/Muhl, Claudia, Mensch-Roboter Interaktion–Status der technischen Entität, Kognitive (Des) Orientierung und Emergenzfunktion des Dritten, in: *Muster Und Verläufe Der Mensch-Technik-Interaktivität*. Working paper: TUTS-WP-2-2012, Berlin 2012.

Deimel, Raphael, Reactive Interaction Through Body Motion and the Phase-State-Machine, in: *IEEE/RSJ International Conference on Intelligent Robots and Systems (IROS)*, Macau 2019, 6383–6390.

Fortunati, Leopoldina/Sorrentino, Alessandra/Fiorini, Laura/Cavallo, Filippo, The Rise of the Roboid, in: *International Journal of Social Robotics*, 13 (6/2021), 1457–1471.

Garfinkel, Harold, *Studies in ethnomethodology*, New Jersey 1967.

Goffman, Erving, *Behavior in Public Places—Notes on the Social Organization of Gatherings*, 1966.

Haus, Juliane, *Das ökonomische Laboratop: Eine soziologische Ethnographie des wirtschaftswissenschaftlichen Experimentierens*, Wiesbaden 2021.

Häussling, Roger, Video analysis with a four-level interaction concept: A network-
based concept of human-robot interaction, in: Ulrike Kissmann (ed.), *Video interaction analysis: Methods and methodology*, Frankfurt a. M. 2009, 107–31.

Heath, Christian/Hindmarsh, Jon, Analysing Interaction: Video, Ethnography and
Situated Conduct, in: *Qualitative Research in Action* (2002), 99–121.

Heath, Christian/Hindmarsh, Jon/Luff, Paul, *Video in Qualitative Research: Analysing
Social Interaction in Everyday Life*, London 2010.

Hornecker, Eva/Bischof, Andreas/Graf, Philipp/Franzkowiak, Lena/Krüger, Nor-
bert, The Interactive Enactment of Care Technologies and its Implications for
Human-Robot-Interaction in Care, in: *Proceedings of the 11th Nordic Conference on
Human-Computer Interaction: Shaping Experiences, Shaping Society*, 2020, 1–11.

Jefferson, Gail, Glossary of transcript symbols with an introduction, in: Gene H.
Lerner (ed.), *Conversation Analysis: Studies from the First Generation*, Amsterdam
2004, 13–31.

Katz, Jack, Analytic Induction, in: Neil J. Smelser/Paul B. Baltes (eds.), *International
Encyclopedia of the Social & Behavioral Sciences*, Oxford 2001, 480–484.

Kendon, Adam, The F-formation system: The spatial organization of social encounters, in: *Man-Environment Systems* 6 (1/1976), 291–296.

Knoblauch, Hubert, Fokussierte Ethnographie: Soziologie, Ethnologie und die neue
Welle der Ethnographie, in: Sozialer Sinn 2 (1/2001), 123–141.

Knoblauch, Hubert/Tuma, René, Videography: An Interpretative Approach to Video-
Recorded Micro-Social Interaction, in: Eric Margolis/Luc Pauwels (eds.), *The
SAGE Handbook of Visual Research Methods*, London 2011, 414–430.

Knorr Cetina, Karin, Sociality with Objects: Social Relations in Postsocial Knowledge
Societies, in: *Theory, Culture & Society* 14 (4/1997), 1–30.

Knorr Cetina, Karin, The Synthetic Situation: Interactionism for a Global World, in:
Symbolic Interaction 32 (1/2009), 61–87.

Krummheuer, Antonia L., Conversation Analysis, Video Recordings, and Human-
Computer Interchanges, in: Ulrike Kissmann (ed.), *Video interaction analysis:
Methods and methodology*, Frankfurt a. M., Berlin, Bern, Wien 2009, 59–83.

Krummheuer, Antonia L./Rehm, Matthias/Rodil, Kasper, Triadic Human-Robot Interaction. Distributed Agency and Memory in Robot Assisted Interactions, in:
*Companion of the 2020 ACM/IEEE International Conference on Human-Robot Interac-
tion*, 2020, 317–319.

Küster, Dennis/Swiderska, Aleksandra/Gunkel, David, I saw it on YouTube! How online videos shape perceptions of mind, morality, and fears about robots, in: *New
Media & Society* 23 (11/2021), 3312–3331.

Latour, Bruno, Give me a laboratory and I will move the world application, in: Karin Knorr Cetina/Michael Mulkay (eds.), *Science observed: Perspectives on the social study of science*, London 1983, 141–170.

Latour, Bruno, *Reassembling the social: An introduction to Actor-Network-Theory* (1st edition), New York 2005.

Laurier, Eric, The panel show: Further experiments with graphic transcripts and vignettes, in: *Social Interaction. Video-Based Studies of Human Sociality* 2 (1/2019).

Lipp, Benjamin (2019). *Interfacing RobotCare. On the Techno-Politics of Innovation*, URL: http://nbn-resolving.de/urn/resolver.pl?urn:nbn:de:bvb:91-diss-20190624-1472757-1-8 [last accessed: August 15, 2023].

Mondada, Lorenza, Multiple Temporalities of Language and Body in Interaction: Challenges for Transcribing Multimodality, in: *Research on Language and Social Interaction* 51 (1/2018), 85–106.

Moon, Ajung/Troniak, Daniel M./Gleeson, Brian/Pan, Matthew K. X. J./Zheng, Minhua/Blumer, Benjamin A./MacLean, Karon/Croft, Elizabeth A., Meet me where I'm gazing: How shared attention gaze affects human-robot handover timing, in: *Proceedings of the 2014 ACM/IEEE International Conference on Human-Robot Interaction*, 2014, 334–341.

Muhle, Florian, Begegnungen mit Nadine. Probleme der "Interaktion" mit einem humanoiden Roboter, in: Angelika Proferl/Michaela Pfadenhauer (eds.), *Wissensrelationen: Beiträge und Debatten zum 2. Sektionskongress der Wissenssoziologie* (1st edition), Weinheim, Basel 2018, 499–511.

Muhle, Florian, Humanoide Roboter als ,technische Adressen': Zur Rekonstruktion einer Mensch-Roboter-Begegnung im Museum, in: *Sozialer Sinn* 20 (1/2019), 85–128.

Muhle, Florian, Roboter in der Sozialwelt. Überlegungen und Einsichten zum Subjektstatus humanoider Roboter, in: M. Schetsche/A. Anton (eds.), *Intersoziologie: Menschliche und nichtmenschliche Akteure in der Sozialwelt*, Weinheim, Basel 2021, 128–142.

Mumm, Jonathan/Mutlu, Bilge, Human-robot proxemics: Physical and psychological distancing in human-robot interaction, in: *Proceedings of the 6th International Conference on Human-Robot Interaction—HRI'11*, 2011, 331–338.

Pan, Matthew K. X. J./Croft, Elizabeth A./Niemeyer, Günter, Exploration of geometry and forces occurring within human-to-robot handovers, in: *2018 IEEE Haptics Symposium (HAPTICS)*, San Francisco 2018, 327–333.

Pfadenhauer, Michaela/Dukat, Christoph, Robot Caregiver or Robot-Supported Caregiving?. The Performative Deployment of the Social Robot PARO in Dementia Care, in: *International Journal of Social Robotics* 7 (3/2015), 393–406.

Pitsch, Karola, Answering a robot's questions. Participation dynamics of adult-child-groups in encounters with a museum guide robot, in: *Réseaux* 220–221 (2/2020), 113–150.

Pitsch, Karola/Gehle, Raphaela/Dankert, Timo/Wrede, Sebastian, Interactional Dynamics in User Groups: Answering a Robot's Question in Adult-Child Constellations, in: *Proceedings of the 5th International Conference on Human Agent Interaction*, 2017, 393–397.

Rammert, Werner, Where the Action is: Distributed Agency between Humans, Machines, and Programs, in: Uwe Seifert/Jin H. Kim/Anthony Moore (eds.), *Paradoxes of Interactivity* (1st edition), Bielefeld 2015, 62–91.

Schegloff, Emanuel, The Relevance of Repair to Syntax-for-Conversation, in: Talmy Givón (ed.), *Discourse and Syntax* (Vol. 12), Boston 1979, 261–286.

Schutz, Alfred, Common-Sense and Scientific Interpretation of Human Action, in: Maurice Natanson (ed.), *Collected Papers I, The Problem of Social Reality*, Dordrecht 1962, 3–47.

Speer, Susan A./Hutchby, Ian, From Ethics to Analytics: Aspects of Participants' Orientations to the Presence and Relevance of Recording Devices, in: *Sociology* 37 (2/2003), 315–337.

Star, Susan L., Simplification in Scientific Work: An Example from Neuroscience Research, in: *Social Studies of Science* 13 (2/1983), 205–228.

Strabala, Kyle W./Lee, Min K./Dragan, Anca/Forlizzi, Jodi/Srinivasa, Siddhartha S./ Cakmak, Maya/Micelli, Vincenzo, Towards Seamless Human-Robot Handovers, in: *Journal of Human-Robot Interaction* 2 (1/2013), 112–132.

Suchman, Lucy, *Human–Machine Reconfigurations: Plans and Situated Actions* (2nd edition), Cambridge 2007.

Tuma, René/Schnettler, Bernt/Knoblauch, Hubert, *Videographie. Einführung in die interpretative Videoanalyse sozialer Situationen*, Wiesbaden 2013.

Veling, Louise/McGinn, Conor, Qualitative Research in HRI: A Review and Taxonomy, in: *International Journal of Social Robotics* 13 (2021), 1689–1709.

Weiss, Astrid/Wurhofer, Daniela/Lankes, Michael/Tscheligi, Manfred, Autonomous vs. tele-operated: How people perceive human-robot collaboration with hrp-2, in: *Proceedings of the 4th ACM/IEEE International Conference on Human Robot Interaction—HRI '09*, 2009a, 257–258.

Weiss, Astrid/Bernhaupt, Regina/Schwaiger, D./Altmaninger, Martin/Buchner, Roland/Tscheligi, Manfred, User experience evaluation with a Wizard of Oz approach: Technical and methodological considerations, in: *9th IEEE-RAS International Conference on Humanoid Robots*, Paris 2009b, 303–308.

Woods, Sarah N./Walters, Michael L./Koay, Kheng L./Dautenhahn, Kerstin, Methodological Issues in HRI: A Comparison of Live and Video-Based Methods in Robot to Human Approach Direction Trials, in: *ROMAN 2006—The 15th IEEE International Symposium on Robot and Human Interactive Communication*, Hatfield 2006, 51–58.

Studying Interaction Indirectly
The Relevance of Secondary Data for Studying Human–Robot Interaction Empirically

Dafna Burema

Abstract *This essay discusses the role of primary and secondary data when empirically studying human–robot interaction (HRI). To understand what type of data to sample, two issues need to be considered: gatekeeping and heterogeneity. Regarding gatekeeping, it is argued that gaining access to the field is difficult for researchers who do not have the financial or symbolic capital needed to study HRI with primary data. Secondary data provide easier access for scholars who are unable to access robots in naturalistic or experimental settings. Furthermore, the field's heterogeneity is suited to case study research. Although this allows for context-specific (i.e., local and temporal) findings, achieving generalizability could be difficult. To move beyond locality and temporality, secondary data enable researchers to expand their scope of study. Whereas primary data allow for the direct study of interactions between humans and robots, representational data of interactions could offer a viable alternative to overcome the issues of gatekeeping and heterogeneity.*

1. Introduction

Using secondary data when studying human–robot interaction (HRI) might sound counterintuitive at first. After all, the main interest is the interaction between the human and the robot, which is directly observable. Nonetheless, this chapter argues that there are pragmatic and theoretical reasons to use secondary data for empirical analyses of HRI[1]. This essay starts by explaining how secondary data, compared to

---

[1]   This text draws on and (re)uses material from my PhD dissertation: Burema, D. (2021). Engineering elder care: An analysis of conceptual premises and biases of social robots in elder care (Doctoral dissertation, Universität Bremen). This PhD thesis was conducted at the Bremen International Graduate School of Social Sciences (BIGSSS), University of Bremen, Jacobs University, which was funded by the European Union's Horizon 2020 research and innovation program under the Marie Skłodowska-Curie grant agreement No 713639.

primary data, overcome the pragmatic issue of gaining access to the field. It then elaborates on theoretical arguments for using secondary data concerning the heterogenous character of the field of HRI. This essay ends with a discussion of the roles of primary and secondary data in generating general versus specific findings, especially in relation to experiments and ethnographies. A particular focus in this latter discussion is how the use of secondary data, compared with primary data, relates to the sample and the research scope.

2. Secondary Data and Field Access

Studying HRI empirically requires both humans and robots. Unfortunately, for some researchers, it is difficult to acquire robots due to gatekeeping. Gatekeeping refers to gaining access to the field,[2] which, in the case of HRI, requires getting access to a robot. This difficulty of gaining access to the field has to do with how expensive the technological artifact is, as well as the requirement of symbolic capital (e.g., network-based or institutional access). Indeed, studying HRI requires that researchers have resources: the financial resources to purchase robots, or a network that enables access to robots. This is similar to the phenomenon of the digital divide,[3] as some researchers and institutions have easier access to robots than others, although one could argue that robots are even more difficult to access than other new technologies. The difficulty of accessing robots could especially hold true for early-stage career researchers who often work on time-restricted projects with a limited budget and a small network. Additionally, researchers and institutions from the global south could also have difficulties accessing such technologies for similar reasons.

Although this essay does not provide solutions to increase financial and symbolic capital among scholars who otherwise cannot access the field of HRI, one way to study HRI with relatively low entry level access is by using secondary data. Researchers could, for instance, sample cases of robots in the academic community by doing a meta-analysis or systematic review, collect data from engineering companies (e.g., their white papers, press releases, etc.), or approach relevant actors in the field of HRI, such as roboticists to, for example, understand how they perceive and experience technology construction. Moreover, researchers could distinguish secondary data as online data, that is, readily available data, or as data acquired from

---

2   E.g., Tom Clark, Gaining and maintaining access: Exploring the mechanisms that support and challenge the relationship between gatekeepers and researchers, in: *Qualitative Social Work* 10 (2011), 485–502.

3   E.g., Jan van Dijk, Digital divide research, achievements and shortcomings, in: *Poetics* 34 (2006), 221–235.

old discussions in the social sciences about "the general" and "the specific," as will be explained next.

4. The General and the Specific

Since the field is very heterogenous in terms of its robots, users, and wider contexts and because it will, arguably, continue to diversify due to the dynamism of computer science and engineering, comparability between cases of HRI is fuzzy. This, together with the logistical constraints of robots being expensive stimuli to study and the necessity of researcher access to specific institutional contexts in which to study HRI, limits scholars' opportunities to expand on different cases. Unsurprisingly then, many studies in the field of HRI are case studies,[10] which sparks the question of how to obtain generalizable findings in HRI.

Before elaborating on that statement, "generalizability" in this essay is the capacity to make statements about a bigger population as a result of studying more than one or two robots comprising the sample. This paper does not refer to generalizability in terms of analytical methods. For instance, grounded theory has the purpose of establishing a general theory that, by default, then embeds some form of generalizability.[11] Although grounded theory might enable researchers to generate theories that travel beyond their research site, an analytical method alone does not provide a solution for understanding HRI with more than just one robot. Similarly, an experiment conducted with one robot stimulus might explain something about the mechanisms between variables, while neglecting how well those mechanisms translate to other robots and their HRI.

The question then remains: In a field as heterogenous as HRI, are generalizations necessary, or is it sufficient to limit empirical studies to one or two cases? Indeed, it could be argued that researchers must cater to the case sensitive nature of HRI—a field in which one size simply does not fit all. A different sample would simply lead to different outcomes, mirroring the diverse landscape of HRI. Papers must then explicitly acknowledge the issues of temporality and local cultures. On the other hand, if the aim is to generalize to a bigger population, then the current research methods need to be re-examined to devise how the issues of temporality and contextuality might be overcome. Specifically, the roles of primary and secondary data ought to be revisited in relation to ethnographies and experiments.

---

10   Cathrine Hasse, The multi-variation approach: Cross-case analysis of ethnographic fieldwork, in: Journal of Behavioral Robotics 10 (2019), 219–227.

11   Keith Taber, Case studies and generalizability: Grounded theory and research in science education, in: *International journal of science education* 22 (2000), 469–487.

5. Ethnographies and Experiments in HRI

When studying HRI as such, that is, with direct observations, two methods are typically used in the field: ethnographies[12] and experiments.[13] However, both methods have specific qualities that often result in studies that involve one or two robots in the research design, as will be explained next.

Ethnographies allow for the study of humans and nonhumans in naturalistic settings.[14] HRI, in this case, is not studied in the lab but rather in "the real world." The ethnographer's role, among others (depending on the exact research design), is to be an observer and/or a participant. This allows for the study of HRI through direct observations, which are primary data. However, ethnographies are often limited in their scope, albeit rich in depth. Ethnographers spend months trying to understand phenomena, which allows for so-called thick descriptions. Unsurprisingly, ethnographies usually involve case studies and thus focus on "the specific" rather than "the general."

In relation to ethnographic research, the anthropologist Hasse[15] has suggested an approach that might increase the generalizability of ethnographic findings. Ethnographers usually study their subject in-depth for a long period of time, ultimately limiting their research to a particular case. Hasse[16] has argued for a multi-variation approach that entails trying to strategically diversify ethnographic cases. In so doing, the researcher still relies on qualitative research methods but is also able to compare cases of social robots. This comparison enables researchers to look beyond one specific case and might even allow them to speak about generalizable patterns. Continuing Hasse's[17] argument, this would especially hold true if the cases are carefully and purposefully selected on the basis of how well they represent the overall population under study. For instance, if a researcher was looking for a typical case of a social robot, the criterionof its prevalence in society, indicated by, for example, sales of the robot, number of users, or media coverage, could be used to determine the case's validity. Taking the example of social robots in elder

---

12   E.g., Michaela Pfadenhauer/Christoph Dukat, Robot caregiver or robot-supported caregiving?, in: *International Journal of Social Robotics* 7 (2015), 393–406.

13   E.g., Rosemarijn Looije/Mark Neerincx/Fokie Cnossen, Persuasive robotic assistant for health self-management of older adults: Design and evaluation of social behaviors, in: *International Journal of Human-Computer Studies* 68 (2010), 386–397.

14   Mike Crang/Ian Cook, *Doing ethnographies*, London 2007.

15   Hasse, The multi-variation approach: Cross-case analysis of ethnographic fieldwork, 219–227.

16   Hasse, The multi-variation approach: Cross-case analysis of ethnographic fieldwork, 219–227.

17   Hasse, The multi-variation approach: Cross-case analysis of ethnographic fieldwork, 219–227.

care, a researcher could hypothetically select Paro, Care-o-bot, or Giraff as cases, to name a few. These robots represent typical cases, but examining them closely for selection reveals radical differences in their hardware and anthropomorphic appearances (e.g., zoomorphic vs. mechanical, short vs. tall, and more), as well as in their objectives (i.e., therapy, assistance, or communication). Having made careful selections with consideration to the study's research design and aim, a researcher can conduct qualitative research and attempt to establish patterns across the field by diversifying the cases.

Nonetheless, it could be argued that one of the reasons numerous user case studies have emerged is due to the high cost of the technological artifact, which makes it simply easier for researchers to work with technology that is readily available at, for example, their university or other affiliated institutions nearby. This relates to the earlier argument of gatekeeping and field access. Hasse's[18] approach requires that ethnographers have a lot of resources: the financial resources to purchase robots, membership in a network that provides access to various robots, and more. Admittedly, although understanding general HRI patterns based on qualitative user studies through, for instance, ethnography would undoubtedly advance the field further, using an ethnographic multi-variation approach is unfeasible for some researchers due to the financial and logistical constrains.

Another research method that enables the study of HRI using primary data, albeit from a completely different epistemological tradition, is experimentation. In experiments, HRI is typically studied with a robot as a stimulus. Although experiments allow for the study of variables that travel beyond the lab, the stimuli used in the field of HRI are typically limited to no more than two robots. This not only has to do with access to the field but also with the complexity of the research design when increasing the number of stimuli, so even if a series of experiments involving different robots were to be conducted, covering cross-case variance would not be a sustainable approach. Indeed, most often, the stimuli are limited to two robots at the most,[19,20] which does not say much about "HRI" as a whole. To give an example, think of experiments such as Heerink et al.'s study, which tested the social abilities of the iCat (an extraverted, socially expressive iCat robot versus an introverted iCat robot that had not been operationalized to be socially expressive) in relation to user acceptance.[21] It could be hypothesized that using a different robot, with different operationalizations of "extraversion" (e.g., in its embodiment, movement, modalities, and other features) would lead to different conclusions. The "mechanism" of

---

18   Hasse, The multi-variation approach, 219–227.
19   E.g., Perugia et al., Modelling engagement in dementia through behaviour, 1112–1117.
20   E.g., Damholdt et al., Attitudinal change in elderly citizens toward social robots, 1–13.
21   Heerink et al., Assessing acceptance of assistive social agent technology by older adults, 361–375.

relating extraversion to user acceptance is therefore limited to a specific robot (in the example, the iCat) and will not necessarily travel across time and contexts.

Experiments are similar to ethnographies regarding the type of observation: Both allow for the direct study of unfolding interactions. Studying "HRI," as such, is typically limited to one of these two methods. Applying other research methods from the social sciences to HRI research leads to the use of secondary data. Although secondary data cannot enable the direct study of interactions, the use of secondary data might enable researchers to extend the scope of studying HRI, as will be explained next.

6. Secondary Data and the Research Scope

The use of secondary data opens up the possibility of sampling many cases since the logistical and research design constraints differ when utilizing secondary data versus primary data. A researcher could, for instance, sample different robotic content by studying many documents (such as white papers or literature reviews on HRI), interviewing several stakeholders from different robot companies, or questioning users of different robots. When using primary data, on the contrary, the researcher typically has to rely on one or two robots that serve as the foundational base for studying HRI using methods such as ethnographic observation or experimentation. Hence, the scope for studying HRI is interlinked with the nature of the data.

Regarding heterogeneity, the issues of temporality and cultural context were previously mentioned as necessary to consider when studying HRI as these have implications for a study's generality or specificity. If a researcher is aiming to broaden the scope and attempt to overcome the contextual relevance that typically marks HRI studies, they could strategically sample documents from different cultural contexts. A meta-analysis or a systematic review could show how HRI unfolds in different cultural settings to ascertain the situatedness or generality of certain findings. Regarding the issue of temporality, studying a variety of cases allows for an understanding of stable elements. Even though robots change, aggregating many different cases allows the researcher to make assumptions about *generations* of robots, meaning that in response to disruptive events, the research only needs to updated if there is reason to believe that new *types* of social robots have emerged, thereby partially overcoming the issue of temporality in the field. The moving target would still be moving, but its pace is easier to keep up with; as opposed to understanding HRI in case-by-case scenarios, general patterns and typologies can be established before new disruptive technologies emerge.

This paper's author conducted an exemplar study aimed at understanding HRI in its broadest sense.[22] The study sought to understand the meaning of "social" in the term "social robot." To achieve that objective, the research sampled 96 academic publications on HRI and analyzed how scholars operationalized social robots for elder care. Many different social robots were analyzed, resulting in a typology of four types of social robots for elder care. How well this robot typology will travel over time with, for example, the emergence of new generations of social robots for use in elder care, remains to be seen. However, robots that currently fit the paradigm can be placed in the typology, introducing an element of stability in the field of HRI. Another study[23] examined HRI scholars' ideas about older adults as prospective users, using the same sample of 96 academic publications on HRI.By taking a bird's eye perspective, the author was able to understand how *systemic* bias in HRI shapes the field and potentially its robots. While not discounting the relevance of studying bias in context with, for example, case studies, sampling a large number of HRI cases led to the understanding that bias is prevalent across the field rather than localized to one case.

The first example given above was a sociological study that aimed to create a robot typology, and the second study was rooted in normative research. In other words, studies differ in terms of the theoretical background and aims. Regardless of what those might be from study to study, secondary data allow scholars to expand the scope of HRI research, thereby enabling an understanding of issues that are *typical, systemic, or recurring* in HRI, that is, focusing on "the stable" instead of "the dynamic." However, researchers must address the biggest drawback of representational data: They do not reflect HRI in "the real world." If the research focus remains on collecting primary data, the researcher could try to implement Hasse's approach[24] of case variation and pay particular attention to selecting cases for the specific research purpose(s). This has proven to be an effective way of understanding HRI directly, and purposeful sampling might enable the researcher to understand HRI in such a way as to produce tentative generalizations. However, given some researchers' lack of resources (e.g., network, financing, organizational constraints, and others), scholars must sometimes resort to more creative solutions to foster generalizability.

---

22   Dafna Burema, Engineering elder care: An analysis of conceptual premises and biases of social robots in elder care, Bremen 2021.

23   Dafna Burema, A critical analysis of the representations of older adults in the field of human–robot interaction, in: AI & SOCIETY 37 (2/2022), 455–465.

24   Hasse, The multi-variation approach, 219–227.

7. Discussion

This essay discussed some key characteristics of secondary data for studying HRI. It has argued that secondary data could partially overcome the issues of gatekeeping and heterogeneity. Concerning the former, similar to the "haves" and "have-nots" that characterize the digital divide, robots are expensive technologies that can be accessed by those who have financial or symbolic capital. Focusing on primary data restricts access to the field since direct observations, by definition, require access to robots. Alternatively, secondary data could allow the "have-nots" to make academic contributions to the field. Such research would not answer certain questions about HRI "performance" or "practices," but by focusing on representations instead, it would open up an otherwise hard to access field.

HRI's heterogeneity taps into discussions about context-specific versus generalizable findings. Indeed, the field of HRI is very diverse, with its different machines, users, and cultural contexts that will, arguably, only diversify over time due to the dynamism of the disciplines of engineering and computer science. This necessitates addressing empirical findings' temporal and local significance. The field's heterogeneity opens up the possibility of conducting case studies but simultaneously makes it difficult to go beyond cases and explore bigger populations and samples. Secondary data, as an alternative to primary data, facilitate the expansion of the empirical research scope. Using primary data often involves methods such as experiments and ethnographies that allow for direct observations of HRI. However, such methods typically do not allow for the study of many robots due to the methods' inherent characteristics. With secondary data, however, other research methods such as systematic reviews, meta-analyses, and questionnaires create possibilities for studying HRI in a broader scope.

It should be noted that one type of data is not necessarily "better" than the other; that is, secondary data are not "inferior" to primary data or vice versa. Although, the "interaction component" of HRI is not studied directly with secondary data, the use of such data could be an option for those who have trouble accessing a research site or want to expand their sample.

Bibliography

Burema, Dafna, Engineering elder care: An analysis of conceptual premises and biases of social robots in elder care, Bremen 2021.

Burema, Dafna, A critical analysis of the representations of older adults in the field of human–robot interaction, in: AI & SOCIETY 37 (2/2022), 455–465.

Clark, Tom, Gaining and maintaining access: Exploring the mechanisms that support and challenge the relationship between gatekeepers and researchers, in: *Qualitative Social Work* 10 (2011), 485–502.

Crang, Mike/Cook, Ian, *Doing ethnographies*, London 2007.

Damholdt, Malene F./Nørskov, Marco/Yamazaki, Ryui/Hakli, Raul/Vesterager Hansen, Catharina/Vestergaard, Christina/Seibt, Johanna, Attitudinal Change in Elderly Citizens Toward Social Robots: The Role of Personality Traits and Beliefs About Robot Functionality, in: *Front. Psychol.* 6 (2015), 1701.

Given-Wilson, Thomas/Legay, Axel/Sedwards, Sean, Information security, privacy, and trust in social robotic assistants for older adults, in: Theo Tryfonas (ed.): *International Conference on Human Aspects of Information Security, Privacy, and Trust*, Cham 2017, 90–109.

Hasse, Cathrine, The multi-variation approach: Cross-case analysis of ethnographic fieldwork, in: *Journal of Behavioral Robotics* 10 (2019), 219–227.

Heerink, Marcel/Kröse, Ben/Evers, Vanessa/Wielinga, Bob, Assessing acceptance of assistive social agent technology by older adults: the almere model, in: *International journal of social robotics* 2 (2010), 361–375.

Kress, Gunther/van Leeuwen, Theo, Colour as a semiotic mode: notes for a grammar of colour, in: *Visual Communication* 1 (2002), 343–368.

Looije, Rosemarijn/Neerincx, Mark/Cnossen, Fokie, Persuasive robotic assistant for health self-management of older adults: Design and evaluation of social behaviors, in: *International Journal of Human-Computer Studies* 68 (2010), 386–397.

Nussey, Sam, SoftBank shrinks robotics business, stops Pepper production- sources (29.06.2021), URL: https://www.reuters.com/technology/exclusive-softbank-s hrinks-robotics-business-stops-pepper-production-sources-2021-06-28/ [last accessed August 15, 2023].

Perugia, Giulia/Diaz Doladeras, Marta/Mallofré, Andreu C./Rauterberg, Matthias/ Barakova, Emilia, Modelling engagement in dementia through behaviour. Contribution for socially interactive robotics, *2017 International Conference on Rehabilitation Robotics (ICORR)*, London 2017, 1112–1117.

Pfadenhauer, Michaela/Dukat, Christoph, Robot caregiver or robot-supported caregiving?, in: *International Journal of Social Robotics* 7 (2015), 393–406.

Taber, Keith, Case studies and generalizability: Grounded theory and research in science education, in: *International journal of science education* 22 (2000), 469–487.

Van Dijk, Jan, Digital divide research, achievements and shortcomings, in: *Poetics* 34 (2006), 221–235.

Conclusion

On the Specifics of Contemporary Forms and Problems of Human–Machine Communication: Concluding Remarks

Florian Muhle, Indra Bock

Abstract *This chapter aims to summarize the insights gained from the contributions to this anthology. It is based on the key questions formulated in the introduction to the book: (1) What are adequate methods for investigating communicative AI in (inter-)action? (2) What are forms and characteristics of interaction with communicative AI? (3) How are encounters with communicative AI framed and shaped by institutional settings? (4) How can interaction with communicative AI in different settings be compared? In this way, both the strengths and weaknesses of ethnographically oriented research in the field of human-machine communication become clear. In addition, the chapter highlights the fact that communication with communicative AI still differs significantly from interpersonal communication and produces its own forms. This shows the (technical) challenges that need to be overcome in order to establish communicative AI in private and institutional contexts outside the laboratory.*

1. Introduction

In the introduction to this anthology, we formulated the expectation that the collected contributions would allow for gaining new and deeper insights into the specifics of contemporary forms and problems of human-machine communication (HMC). In particular, we were hoping to find answers to the following questions:

- What are adequate methods for investigating communicative AI in (inter-)action?
- What are forms and characteristics of interaction with communicative AI?
- How are encounters with communicative AI framed and shaped by institutional settings?
- How can interaction with communicative AI in different settings be compared?

To conclude the book, in this chapter we aim at briefly summarizing the extent to which these questions in the chapters have been answered. In this way, we highlight the central insights provided by this anthology as well as the analytical potential of ethnographically oriented social science research on communicative AI 'in the wild'. In addition, however, the challenges that this type of research has to face become also apparent.

2. What are Adequate Methods for Investigating Communicative AI in (Inter-)Action?

As already described in the introduction, research into communicative AI in (inter-)action poses particular challenges. This is because HMC 'in the wild' does not take place in controlled environments, as is the case with laboratory experiments. Instead, communicative AI systems that already have left the laboratories of the technical sciences are embedded in various everyday and institutional contexts and must meet the respective requirements of the different social situations.

What exactly happens in these settings is generally unpredictable and largely characterized by practical routines that can only be verbalized to a very limited extent. It follows from this that research that wants to approach the particularities of HMC 'in the wild' and understand them in detail cannot rely (alone) on standardized research methods or on interview methods commonly used in social research. The former does not allow for open-ended and exploratory research, while the latter can only ascertain what the interviewees are aware of. Therefore, respective approaches only offer limited insight into the problems and peculiarities of human-machine communication *in situ*, which makes other approaches necessary[1].

As the contributions in this anthology have shown, (auto-)ethnographic methods are particularly suitable for such research. After all, ethnography thrives on openly engaging with everyday (and institutional) life and exploring it through observation. The more detailed the research interest, the more it makes sense in this context to use audiovisual methods of data recording, as has been established for many years, particularly in 'focused ethnography'[2]. This is because audiovisual recordings, which were used in almost every study in this anthology, enable the repeated, detailed, and intensive analysis of data, which allows for gaining insights

---

1   Kerstin Dautenhahn, Robots in the Wild. Exploring Human-Robot Interaction in Naturalistic Environments, in: *Interaction Studies* 10 (3/2009), 269–273, see 270.

2   Hubert, Knoblauch, Focused Ethnography, in: *Forum Qualitative Sozialforschung / Forum: Qualitative Social Research*, 6(3/2005), Art. 44, http://nbn-resolving.de/urn:nbn:de:0114-fqs05034 40.

that would hardly be possible with other methods. An example of this is the discovery of specific 'members' methods' for doing 'VUI-speak' in the contribution by *Due and Lüchow*.

It thus becomes clear—as *Maibaum et al.* in particular point out—that the use of audiovisual recordings in research into human-machine communication should not only serve illustrative purposes but can be highly knowledge-generating when used in a systematic and methodologically controlled manner. *Maibaum et al.* provide explicit hints of how such a reflective use can take place. In addition, each study in this book also shows in an exemplary manner that and how audiovisual recordings can be used to answer specific questions in the context of investigating communicative AI in (inter-)action.

Looking at the various contributions, it also becomes clear how the methods are adapted in specific ways in the individual research settings and thus realize the "unique adequacy requirement of methods"[3], as is demanded in the context of ethnomethodological research in particular. For example, *Harth* uses 'mixed reality methods' to trace the interactions between humans and an embodied agent in virtual reality, while *Muhle et al.* combine the analysis of audiovisually recorded data with the analysis of program code to gain a comprehensive view of the architecture-for-interaction of an embodied conversational agent.

Overall, the contributions thus convincingly demonstrate the suitability of ethnographic research for the investigation of communicative AI in (inter-)action. They thus (hopefully) contribute to the establishment of corresponding approaches in this field of research, as they are especially well-suited to delve into the specifics of contemporary forms and challenges of human-machine communication.

3. What are Forms and Characteristics of Interaction with Communicative AI?

From the developers' perspective, the establishment of communicative AI primarily aims to enable intuitive and 'natural' interaction between humans and machines. In contrast to this, the chapters in this anthology show that this goal has clearly not been achieved to date. On the contrary, human-machine communication still differs noticeably from 'regular' interpersonal communication. This holds true in various respects. As *Cuevas-Garcia and O'Donovan* point out in their contribution, the possibilities of HMC in the field of social robotics are still mostly limited to "prescribed forms of interaction", which differ significantly from the normal course of "situated

---

3   Harold Garfinkel/D. Lawrence Wieder, Two incommensurable, asymmetrically alternate technologies of social analysis, in: Graham Watson/Robert Morris Seiler (eds), *Text in Context. Contributions to Ethnomethodology.* New York: Sage 1992, pp.175–206.

actions"[4] in everyday and institutional settings. The public use of social robots is thus based on a 'logic of control' that makes spontaneous and situationally adapted encounters between humans and robots difficult.

Similarly, *Langedijk and Fischer* work out that the use of a drink-service robot in care facilities does not simply fit into everyday care practice, but is associated with considerable requirements and adaptation problems that were not anticipated by the developers. In part – as in the study by *Cuevas-Garcia and O'Donovan* – this is due to the design of the environment in which the robot is used.

In addition, the interface design appears less intuitive than intended by the developers, which makes interaction between humans and robots more difficult. The chapters by *Muhle et al.* and *Harth* show that this does not only apply to individual cases. *Muhle et al.* show the extent to which the design of the interface of the embodied agent they investigated systematically leads to difficulties in operation and interaction. Conversely, *Harth* works out the problem that the agent he investigated is not able to understand non-verbal aspects of communication. These design problems result in asymmetrical communication situations between humans and machines, as the latter do not have the same communicative resources as their human counterparts.

Asymmetries between humans and communicative AI systems can also be seen in the fact that the technical systems are still barely able to participate competently in everyday conversations, as *Lind* as well as *Due and Lüchow* show in their chapters. This is because AI systems still today lack the ability for situated understanding, which systematically gives rise to communication problems that cannot be reliably solved by meta-communication and still constitute a specific feature of contemporary HMC. *Muhle et al.* provide an example of how such communication problems arise when they integrate an analysis of the internal operational processes of an embodied agent into their analysis of a human-machine encounter.

Due to the limitations of machine communication capabilities, unique forms of HMC are established that must be clearly distinguished from human interactions. Examples include the aforementioned special features of 'VUI-speak' outlined by *Due and Lüchow*, but also forms of testing the communicative capabilities of technical systems. These can be seen, for example, in provocations of the system, as described by *Lind* and also discovered in other HMC studies[5].

---

4   Lucy Suchman, *Human-Machine Reconfigurations. Plans and Situated Actions* (2nd edition), Cambridge 2007.

5   Antonia Krummheuer, Herausforderung künstlicher Handlungsträgerschaft. Frotzelattacken in hybriden Austauschprozessen von Menschen und virtuellen Agenten, in: Hajo Greif/Oana Mitrea/Matthias Werner (eds.): *Information und Gesellschaft: Technologien einer sozialen Beziehung.* Wiesbaden: VS Verlag für Sozialwissenschaften 2008, pp. 73–95.

Overall, the contributions in this book show that the development of communicative AI continues to fall short of its goal to enable intuitive, human-like communication. Whether and to what extent this will change in the future remains subject to further research. It is possible that further technological progress will bring HMC closer to its goal to enable communication with machines that is not distinguishable from human communication. Until then, however, the special features and limitations of communication with communicative AI must continue to be examined and systematically described. Ultimately, this can also contribute to the further development of the technical systems, as it brings to light the weaknesses of the systems that were not anticipated by developers.

4. How are Encounters with Communicative AI Framed and Shaped by Institutional Settings?

A special feature of human communication is that it can constantly be adapted to new contexts by the people involved. For example, people who regularly buy something to eat in a supermarket during their lunch break are intuitively able to switch from a 'sales interaction' to 'small talk' and back again during a conversation at the checkout, and then return to their professional role when they are back in the office. This ability to adapt to different situational and institutional contexts can hardly be expected from machines today. Their limited communication skills described above (see section 3) speak against this. It therefore seems necessary to implement AI systems that are used in institutional contexts with typical activities that are relevant in these contexts.

The contributions by *Cuevas-Garcia and O'Donovan* and *Langedijk and Fischer* show how great the challenge already is here to prepare corresponding capabilities for specific and clearly defined application contexts. In *Langedijk's and Fischer's* chapter, this is made clear by the example of the task of serving drinks in a care facility. *Cuevas-Garcia and O'Donovan* demonstrate this by describing the technical problems that arise when carrying out a robot competition. Both chapters thus show very clearly that and how the situational and institutional context prefigures the possibilities and limitations of HMC. In both cases, the possible applications are extremely limited and even serving drinks in a care facility or taking orders in a café prove to be major challenges under real-world conditions 'in the wild'.

At the same time, especially *Lind's* contribution shows that and how humans 'domesticate' communicative AI in a specific way and thus integrate it into the reproduction of specific private or institutional contexts. Once again, here the limited communicative capabilities of the technical systems pose a challenge, and it is not possible to integrate them into corresponding contexts in the same way as humans. However, this does not prevent people from actively using communicative

AI. In Lind's case study, for example, the commonalities between the human family members are emphasized and reinforced when they are 'doing family' by the fact that they stand in stark contrast to the limited communicative abilities of the smart speaker Alexa, which make it almost impossible to socialize with the system.

Again, this shows that it is the humans, who are able to adapt spontaneously and situationally to changing contexts and situations, while 'intelligent' machines lack this ability and hence must be prepared accordingly for specific tasks in predefined contexts. However, as soon as unforeseen events occur in these contexts, this also becomes a major challenge for the technical systems involved.

5. How can Interaction with Communicative AI in Different Settings be Compared?

The contributions in this anthology, as well as the preceding remarks, have shown that overarching insights are possible with regard to the specifics of the contemporary forms and problems of human-machine communication. On the one hand, the existing communicative limitations of the systems, be they robots, agents or smart speakers, become apparent across all cases. On the other hand, it also became clear how difficult it is to establish communicative AI in concrete application scenarios in private or institutional settings.

At the same time, however, it is also clear that, beyond the mentioned general insights, a systematic comparison of communicative AI systems in (inter-)action in different settings is difficult. After all, in the different studies, which served as basis for the chapters of this book, very different systems and very specific application scenarios were examined. In addition, the questions posed by the individual contributions were quite diverse, which also makes comparison difficult in this respect. In this sense, ethnographically oriented research, which aims at the case-oriented, open and unstandardized exploration of human-machine communication in situ, also shows its weaknesses: openness and case-sensitivity simply make comparability difficult.

This problem is addressed in the last chapter, in which *Burema* emphasizes the relevance of secondary analyses, including meta-analyses and systematic reviews. Such analyses can help to sort out the heterogeneity of systems, of research questions and of empirical results. In this way they not only allow for mapping the state of research, but also make clear in which respects comparisons are possible and in which they are difficult. Based on this, research gaps can also be identified.

In this context, the archiving and accessibility of primary data from empirical studies, such as the provision of transcriptions and – where possible – audiovisual data, could also be helpful. While this is already common practice in quantitatively oriented research and is also demanded by funding agencies, corresponding attempts to archive qualitatively generated data are still in their infancy and are associated with far greater difficulties. However, such archives would enable researchers to develop questions specifically geared towards comparison and to investigate these using the data provided. This could gradually reduce the single-case focus of research and pave the way for more comprehensive analyses.

In sum, not only the development of communicative AI, but also its observation and research from a social sciences perspective continues to be associated with major challenges. We believe that the contributions collected in this anthology provide convincing indications of what these challenges are and how they can be addressed. We have no doubt that the need for corresponding research will continue to increase in the coming years as communicative AI systems become more and more integrated into everyday life, both in private and in institutional contexts.

Bibliography

Dautenhahn, Kerstin, Robots in the Wild. Exploring Human-Robot Interaction in Naturalistic Environments, in: *Interaction Studies* 10 (3/2009), 269–273.

Garfinkel, Harold / Wieder, D. Lawrence, Two incommensurable, asymmetrically alternate technologies of social analysis, in: Graham Watson/Robert Morris Seiler (eds), *Text in Context. Contributions to Ethnomethodology*. New York: Sage 1992, 175–206.

Knoblauch, Hubert, Focused Ethnography, in: *Forum Qualitative Sozialforschung / Forum: Qualitative Social Research*, 6(3/2005), Art. 44, http://nbn-resolving.de/urn:nbn:de:0114-fqs0503440.

Krummheuer, Antonia, Herausforderung künstlicher Handlungsträgerschaft. Frotzelattacken in hybriden Austauschprozessen von Menschen und virtuellen Agenten, in: Hajo Greif/Oana Mitrea/Matthias Werner (eds.): *Information und Gesellschaft: Technologien einer sozialen Beziehung*. Wiesbaden: VS Verlag für Sozialwissenschaften 2008, 73–95.

Suchman, Lucy, *Human-Machine Reconfigurations. Plans and Situated Actions* (2nd edition), Cambridge 2007.

Authors

Indra Bock is a Ph.D. student at Bielefeld Graduate School in History and Sociology and a research fellow at Zeppelin University in the 3B Bots Building Bridges project. Her background is in qualitative social research and media sociology. Indra's work surrounds human-robot interaction as well as automated communication in Online Social Networks and its influence on political opinion formation.

Dafna Burema is a postdoctoral researcher at the Institute of Sociology at Technische Universität Berlin, and Science of Intelligence (Research Cluster of Excellence). She has a particular interest in AI ethics, critical theory, and uses qualitative research methods in her work.

Carlos Cuevas-Garcia is a postdoctoral researcher at the Department of Science, Technology and Society of the Technical University of Munich. He studies digital innovation, societal transformations, and interrelations between research policy and collaboration practices.

Brian L. Due is an associate professor in communication at the University of Copenhagen. He studies sociomaterial interactions, mobility and social organizations using video ethnography.

Kerstin Fischer is professor for Language and Technology Interaction at the University of Southern Denmark and director of the Human-Robot Interaction Lab in Sonderborg. Kerstin is senior associate editor of the journal ACM Transactions on Human-Robot Interaction and associate editor of the book series 'Studies in Pragmatics' (Brill). She has published 9 books, 35 journal articles and more than 100 conference papers, in which she brings her background in linguistics, communication and multimodal interaction analysis to the study of behavior change, persuasive technology and human-robot interaction.

Philipp Graf is a sociologist of technology and doctoral candidate at Chemnitz University of Technology. His research focuses on social robotics, especially its use in care and medicine, and qualitative methds.

Jonathan Harth is a sociologist at Witten/Herdecke University and does research on the use of extended reality technologies and sociality under the conditions of artificial intelligence. In addition, he is currently working in the project Theatre of Augmented Realities at the Theater an der Ruhr, Mülheim.

Rosalyn M. Langedijk is a research assistant at the Department of Design, Media and Educational Science at the University of Southern Denmark in Sonderborg. Her research concerns communication design, human-robot interaction, and ethnography. She investigates persuasive dialogs in human-robot interaction in laboratory and real-world settings.

Miriam Lind is leader of the research group „Posthumanist Linguistics. Communicative Practices between Humans, Animals, and Machines" at the European University Viadrina in Frankfurt/Oder. Her research concerns discursive and interactional negotiations of the boundaries between humans and nonhumans, and the ways communication technologies impact human-animal relationships.

Louise Lüchow is a PhD student at the Department of Nordic Studies and Linguistics at the University of Copenhagen. Her research is rooted in ethnomethodology and conversation analysis, and focuses on social interaction, technology, and digital transformation, particularly how visually impaired individuals use AI technology for distributed perception.

Arne Maibaum is a PhD candiate at TU Berlin. His research focuses on the sociology of technology, science and technology studies, and human-robot-interaction.

Henning Mayer is an Account Technology Strategist at Microsoft and a Ph.D. student at the University of Hamburg. His research focuses on human-machine interaction, social robotics and sociometric models of artificial intelligence.

Florian Muhle is professor of Communication Studies with a focus on digital communication at Zeppelin University Friedrichshafen. His research interests cover human-machine communication, automation of communication and the digital transformation of the public sphere.

Cian O'Donovan is a Senior Research Fellow at UCL's Department of Science and Technology Studies. He studies the policies and processes of digital change using social science-led interdisciplinary approaches collaborating with people driving innovation and directly impacted by innovation.

René Tuma is a postdoctoral sociologist based at Technische Universität Berlin. He has published on videography from both methodological and reflexive perspectives. His research interests include sociology of interaction, knowledge and technology. Research areas include policing, violence, internet governance and vernacular methods.

GPSR Authorized Representative: Easy Access System Europe, Mustamäe tee 50, 10621 Tallinn, Estonia, gpsr.requests@easproject.com